T0189345

Studies in Computational Intelligence

Volume 811

Series editor

Janusz Kacprzyk, Polish Academy of Sciences, Warsaw, Poland
e-mail: kacprzyk@ibspan.waw.pl

The series "Studies in Computational Intelligence" (SCI) publishes new developments and advances in the various areas of computational intelligence—quickly and with a high quality. The intent is to cover the theory, applications, and design methods of computational intelligence, as embedded in the fields of engineering, computer science, physics and life sciences, as well as the methodologies behind them. The series contains monographs, lecture notes and edited volumes in computational intelligence spanning the areas of neural networks, connectionist systems, genetic algorithms, evolutionary computation, artificial intelligence, cellular automata, self-organizing systems, soft computing, fuzzy systems, and hybrid intelligent systems. Of particular value to both the contributors and the readership are the short publication timeframe and the world-wide distribution, which enable both wide and rapid dissemination of research output.

The books of this series are submitted to indexing to Web of Science, EI-Compendex, DBLP, SCOPUS, Google Scholar and Springerlink.

More information about this series at http://www.springer.com/series/7092

Seyedali Mirjalili · Jin Song Dong ·
Andrew Lewis
Editors

Nature-Inspired Optimizers

Theories, Literature Reviews and Applications

 Springer

Editors
Seyedali Mirjalili
Institute for Integrated
and Intelligent Systems
Griffith University
Brisbane, QLD, Australia

Andrew Lewis
Institute for Integrated
and Intelligent Systems
Griffith University
Brisbane, QLD, Australia

Jin Song Dong
Institute for Integrated
and Intelligent Systems
Griffith University
Brisbane, QLD, Australia

Department of Computer Science
School of Computing
National University of Singapore
Singapore, Singapore

ISSN 1860-949X   ISSN 1860-9503   (electronic)
Studies in Computational Intelligence
ISBN 978-3-030-12129-7   ISBN 978-3-030-12127-3   (eBook)
https://doi.org/10.1007/978-3-030-12127-3

Library of Congress Control Number: 2018968101

This Springer imprint is published by the registered company Springer Nature Switzerland AG
The registered company address is: Gewerbestrasse 11, 6330 Cham, Switzerland

To our parents

Preface

One of the fastest growing sub-fields of Computational Intelligence and Soft Computing is Evolutionary Computation. This field includes different optimization algorithms that are suitable for solving NP-hard problems for which exact methods are not efficient. Such algorithms mostly use stochastic operators and are gradient-free, which makes them suitable for solving nonlinear problems, particularly those for which objectives are noisy, multi-modal, or expensive to evaluate.

The main purpose of this book is to cover the conventional and most recent theories and applications in the area of Evolutionary Algorithms, Swarm Intelligence, and Meta-heuristics. The chapters of this book are organized based on different algorithms in these three classes as follows:

- Ant Colony Optimizer
- Ant Lion Optimizer
- Dragonfly Algorithm
- Genetic Algorithm
- Grey Wolf Optimizer
- Grasshopper Optimization Algorithm
- Multi-Verse Optimizer
- Moth-Flame Optimization Algorithm
- Salp Swarm Algorithm
- Sine Cosine Algorithm
- Whale Optimization Algorithm

Each chapter starts by presenting the inspiration(s) and mathematical model(s) of the algorithm investigated. The performance of each algorithm is then analyzed on several benchmark case studies. The chapters also solve different challenging problems to showcase the application of such techniques in a wide range of fields. The problems solved are in the following areas:

- Path planning
- Training neural networks
- Feature selection

- Image processing
- Computational fluid dynamics
- Hand gesture detection
- Data clustering
- Optimal nonlinear feedback control design
- Machine learning
- Photonics

Brisbane, Australia   Dr. Seyedali Mirjalili
August 2018   Prof. Jin Song Dong
                                                            Dr. Andrew Lewis

Contents

Grasshopper Optimization Algorithm: Theory, Literature Review, and Application in Hand Posture Estimation   107
Shahrzad Saremi, Seyedehzahra Mirjalili, Seyedali Mirjalili
and Jin Song Dong

Multi-verse Optimizer: Theory, Literature Review, and Application in Data Clustering   123
Ibrahim Aljarah, Majdi Mafarja, Ali Asghar Heidari, Hossam Faris
and Seyedali Mirjalili

Contributors

Ibrahim Aljarah King Abdullah II School for Information Technology, The University of Jordan, Amman, Jordan

Jin Song Dong Institute for Integrated and Intelligent Systems, Griffith University, Nathan, Brisbane, QLD, Australia;
Department of Computer Science, School of Computing, National University of Singapore, Singapore, Singapore

Hossam Faris King Abdullah II School for Information Technology, The University of Jordan, Amman, Jordan

Ali Asghar Heidari School of Surveying and Geospatial Engineering, University of Tehran, Tehran, Iran

Andrew Lewis Institute for Integrated and Intelligent Systems, Griffith University, Nathan, Brisbane, QLD, Australia

Majdi Mafarja Department of Computer Science, Faculty of Engineering and Technology, Birzeit University, Birzeit, Palestine

Seyed Hamed Hashemi Mehne Aerospace Research Institute, Tehran, Iran

S. Mirjalili Institute for Integrated and Intelligent Systems, Griffith University, Nathan, Brisbane, QLD, Australia

Seyed Mohammad Mirjalili Department of Electrical and Computer Engineering, Concordia University, Montreal, QC, Canada

Seyedali Mirjalili Institute of Integrated and Intelligent Systems, Griffith University, Nathan, Brisbane, QLD, Australia;
School of Information and Communication Technology, Griffith University, Brisbane, QLD, Australia

Seyedeh Zahra Mirjalili School of Electrical Engineering and Computing, University of Newcastle, Callaghan, NSW, Australia

Ali Safa Sadiq School of Information Technology, Monash University, Bandar Sunway, Malaysia

Shahrzad Saremi Institute for Integrated and Intelligent Systems, Griffith University, Brisbane, QLD, Australia

Introduction to Nature-Inspired Algorithms

Seyedali Mirjalili and Jin Song Dong

Abstract This chapter is an introduction to nature-inspired algorithms. It first discusses the reason why such algorithms have been very popular in the last decade. Then, different classifications of such methods are given. The chapter also includes the algorithms and problems investigated in this book.

In the field of Artificial Intelligence (AI), there is a large number of nature-inspired techniques to solve a wide range of problems. One of the most specialized areas in AI with the most nature-inspired techniques is Computational Intelligence (CI). There is no agreed definition for CI, but a large number of researchers believe it is equivalent to the field of Soft Computing. In both of these fields, the main objective is to develop inexact solutions for NP-complete problems where there is not exact solution that runs in polynomial time.

As an example, the Traveling Salesman Problem (TSP) is worth mentioning here, in which a traveller wants to find the shortest path to visit n cities in a country. As example is shown in Fig. 1 where the objective is to find the shortest path to visit 20 biggest cities in Australia. Assuming that the traveller can visit any city in any order and (s)he can travel directly from one city to any other, all the possible connections between cities are shown in Fig. 2.

This problem seems very simple since there are only 20 cities. However, this chapter shows that it is one of the most challenging NP-hard problems that humankind has ever faced. The size of search space can be calculated using $n!$. This means that

S. Mirjalili (✉) · J. Song Dong
Institute for Integrated and Intelligent Systems, Griffith University, Nathan,
Brisbane, QLD 4111, Australia
e-mail: seyedali.mirjalili@griffithuni.edu.au

J. Song Dong
Department of Computer Science, School of Computing, National University of Singapore,
Singapore, Singapore
e-mail: j.dong@griffith.edu.au

© Springer Nature Switzerland AG 2020
S. Mirjalili et al. (eds.), *Nature-Inspired Optimizers*, Studies in Computational
Intelligence 811, https://doi.org/10.1007/978-3-030-12127-3_1

Fig. 1 Twenty biggest cities
in Australia

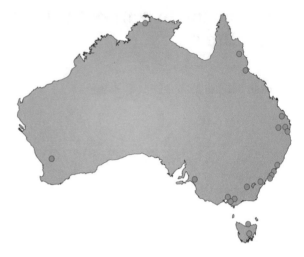

Fig. 2 The search space of
finding the shortest route to
visit 20 cities in Australia

there are there are $2.432902e + 18$ possible tours to search from. If it takes someone to calculate the length of each tour one second, it takes 77,000,000,000 years to search the entire search space, which is nearly 20 times longer than the big bang. Even using a computer is not helpful here. Assuming that we use a computer with 2.0 GHz CPU, which means 2,000,000,000 instructions per second, checking all the possible routes will take nearly 38 years. Adding one city to this problem increases this period of time to nearly 810 years.

The main challenge in such NP-hard problems is the exponential growth in the resources (often time and memory) required to solve them. The growth of $n!$ is even more then exponential as can be seen in Fig. 3. Exact methods are effective for emblems with polynomial time and their performance substantially degrades when solving problems with exponential growth. It should be noted that the problem

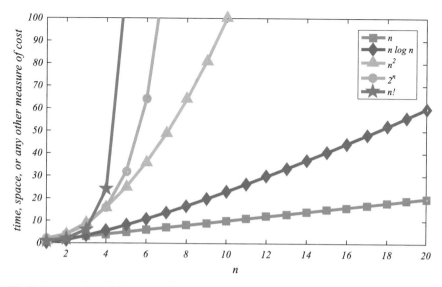

Fig. 3 A comparison of the computational cost of different problems including NP-hard problems: $O(n)$, $O(n \log n)$, $O(n^2)$, $O(2^n)$, and $O(n!)$

mentioned in this chapter has been taken from the field of combinatorial optimization where the variables are discrete. This means that the size of search space is finite. A large number of real-world problems, however, have continuous variables. This means that the search space for these kinds of problems is indefinite. This again substantially increases the difficulty of the problem and consequently for exact methods to solve them.

To solve such exponentially growing problems with either discrete and/or continuous variables, there is no exact solutions that find the best solution in a polynomial time specially when the number of variables is large. In the field of CI or soft computing, computational tools, methods, and algorithms are imprecision tolerant, uncertain, partially true, or using approximations. This makes them unreliable in finding exact solutions, but the literature shows that they can find near optimal (occasionally optimal) solution to NP-hard problems in a reasonable time. For instance, Ant Colony Optimization [1] can find a very short path for the TSP problem with 20 cities in a fraction of a second. With running this algorithm multiple times, it can find the optimal path. So in the worst case, it is still substantially faster than exact methods for this problem. Of course, completely random search methods can be used here too, but they are not systematic, and they perform equivalent to an exhaustive search in the worst case.

Figure 4 shows where soft computing techniques are in the context of search methods. It can be seen that exact methods are the most reliable techniques which will eventually find the solution when resources are not limited. However, they are not effective for NP-hard problems as discussed above. On the other hand, random

Fig. 4 Comparison of a
brute-force (exhaustive)
search,
soft-computing-based
search, and random search

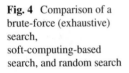

search techniques might be so much faster. The reliability of such methods is the main drawback since they are not systematic and the search is completely random. Soft computing techniques are in the 'middle', in which they incorporate systematic, stochastic, imperfect, and partially true components to find the optimal solutions for problems in reasonable time.

The field of CI and soft computing can be divided into three subfields: Fuzzy Logics, Neural Networks (Machine Learning), and Evolutionary Computation. In the first class, another dimension is given to Boolean logic where something can be partially true or false too. The field of machine learning uses a wide range of techniques, including well-regarded Neural Networks, to give computers the ability to learn. Finally, the field of evolutionary computation includes problem solving techniques (algorithms) inspired from nature.

This book contributes to the third area, evolutionary computation. This sub-field investigates evolutionary concepts in nature and simulates them in computer to solve problems. It should be noted that despite the word 'evolutionary' in the name of this field, other nature-inspired algorithms are also lie within this class.

The main purpose of this book is to cover the conventional and most recent theories and applications in the area of evolutionary algorithms, swarm intelligence, and meta-heuristics as shown in Fig. 5.

The chapters of this book are organized based on different algorithms in the three above-mentioned classes. The chapters are as follows:

- Ant Colony Optimizer
- Ant Lion Optimizer

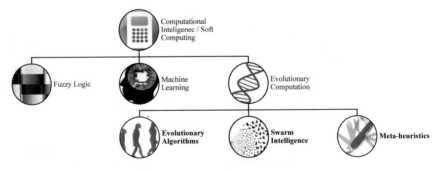

Fig. 5 The main areas that this book contributes to: Evolutionary Algorithms, Swarm Intelligence, and Meta-heuristics

- Dragonfly Algorithm
- Genetic Algorithm
- Grey Wolf Optimizer
- Grasshopper Optimization Algorithm
- Multi-verse Optimizer
- Moth-Flame Optimization Algorithm
- Salp Swarm Algorithm
- Sine Cosine Algorithm
- Whale Optimization Algorithm

Each chapter starts with presenting the inspiration(s) and mathematical model(s) of the algorithm investigated. The performance of each algorithm is then analyzed on several benchmark case studies. All chapters also solve a different problem to showcase the application of such techniques in a wide range of fields. The problems solved are in the following areas:

- Path planning
- Training Neural Networks
- Feature Selection
- Image processing
- Computational Fluid Dynamics
- Hand gesture detection
- Data clustering
- Optimal nonlinear feedback control design
- Machine learning
- Photonics

Reference

1. Dorigo, M., & Birattari, M. (2011). Ant colony optimization. In *Encyclopedia of machine learning* (pp. 36–39). Boston, MA: Springer.

Ant Colony Optimizer: Theory, Literature Review, and Application in AUV Path Planning

Seyedali Mirjalili, Jin Song Dong and Andrew Lewis

Abstract This chapter starts with the inspiration and main mechanisms of one of the most well-regarded combinatorial optimization algorithms called Ant Colony Optimizer (ACO). This algorithm is then employed to find the optimal path for an AUV. In fact, the problem investigated is a real-world application of the Traveling Salesman Problem (TSP).

1 Introduction

Ant Colony Optimization (ACO) [1, 2] is one of the well-known swarm intelligence techniques in the literature. The original version of the algorithm is suitable for solving combinatorial optimization problems (*e.g.,* vehicle routing and scheduling). However, they have been a lot of modification in this algorithms that make it capable of solving a wide rage of problems these days. Despite the similarity of this algorithm to other swarm intelligence techniques in terms of employing a set of solutions and stochastic nature, the inspiration of this algorithm is unique. This chapter presents the inspirations and mathematical models of this algorithm for solving both combinatorial and continuous problems.

S. Mirjalili (✉) · J. Song Dong · A. Lewis
Institute for Integrated and Intelligent Systems, Griffith University, Nathan,
Brisbane, QLD 4111, Australia
e-mail: seyedali.mirjalili@griffithuni.edu.au

J. Song Dong
Department of Computer Science, School of Computing, National University of Singapore,
Singapore, Singapore
e-mail: j.dong@griffith.edu.au

A. Lewis
e-mail: a.lewis@griffith.edu.au

© Springer Nature Switzerland AG 2020
S. Mirjalili et al. (eds.), *Nature-Inspired Optimizers*, Studies in Computational
Intelligence 811, https://doi.org/10.1007/978-3-030-12127-3_2

2 Inspiration

The main inspiration of the ACO algorithm comes from stigmergy [3]. This refers to the interaction and coordination of organisms in nature by modifying the environment. In stigmergy, the trace of an action done by an organism stimulates subsequent actions by the same or other organisms. For instance, in a termite colony, one termite might role a ball of mud and left it next to a hole. Another termite identifies the mud and uses it to fix the hole. In nature, this causes complex and decentralized intelligence without planning and direct communication.

One of the most evident stigmergy can be found in the behavior of ants in an ant colony when foraging food. Normally, foraging is an optimization task where organisms try to achieve maximum food source by consuming minimum energy. In an ant colony, this can be achieved by finding the closest path from the nest to any food source. In nature, ants solve this problem using a simple algorithm [4].

Ants communicate using pheromone. If an ant does not sense pheromone, it randomly searches for food sources in an environment. After finding a food source, it starts marking the path with pheromone. The amount of the pheromone highly depends on the quality and quantity of the food source. The more and the better the source of food, the stronger and concentrated th pheromone is. This attracts other ants towards the food source. An ant might even abandon its own route to a food source in case of sensing a route with stronger pheromone. What causes finding the shortest path is the attraction of more ants and depositing more amount of pheromone over time. Pheromone also vaporizes, so a short part has a higher chance of being re-deposited with pheromone as compare to a long path. This means that the pheromone on the closets route become more concentrated as more ants get attracted to the strongest pheromone. This is conceptually modeled in Fig. 1.

This figure shows that the upper path attracts less ants since the amount of pheromone is less. The lower path is more crowded since it is a bit shorter than the upper one and the pheromone vaporizes slower. Finally, the path in the middle attracts most of the ants since the impact of vaporization is reduced since many ants top up the path with pheromone. Over time, more and more ants get attracted to the shortest path and at some point, no ant chooses the long routes since the pheromone completely vaporizes before it gets refilled by other ants.

In 1996, Marco Dorigo inspired from this behavior and proposed and optimization algorithm called ACO. The first theory of optimization using ant models is called Ant System proposed in 1992 [5]. There are also other seminal works in this area such as Max-Min Ant System proposed in 1996 [6] and ACO proposed in 1997 [7]. The mathematical model and pseudo-code of the ACO algorithm is presented in the following sections [5].

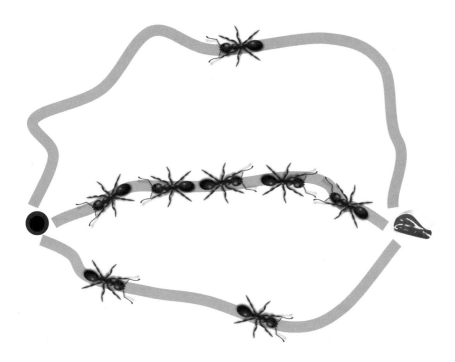

Fig. 1 Three routes from an nest to a food source. The amount of pheromone deposited on a route is of the highest on the closest path. The of pheromone decreases proportionally to the length of the path. While ants add pheromone to the paths towards the food source, vaporization occurs. The period of time that an ant tops up pheromone before it vaporizes is higher inversely proportional to the length of the path. This means that the pheromone on the closest route becomes more concentrated as more ants get attracted to the strongest pheromone

3 Mathematical Model

The ACO algorithm is suitable for combinatorial optimization problems [8]. In such problems, the variables are selected from a finite set of values. The objective is to find an optimal set of values from a finite set to maximize or minimize an objective. Most of the combinatorial problems are NP-hard, so the performance of deterministic algorithms significantly degrades when increasing the scale of the problem. Similar to other heuristics and meta-heuristics, ACO employs a mechanism to avoid searching the entire search space and only focus on promising regions.

ACO has been equipped with three main operators to solve combinatorial optimization problems: construction of artificial ants, pheromone update, and daemon (optional). The pseudo-code of the ACO algorithm and the sequence of the main phases are given in Fig. 2. The concepts and mathematical models of construction, pheromone update, and daemon phases are discussed in the following sections.

Fig. 2 The pseudo code of
the ACO algorithm

```
Initialization
while the end condition is not satisfied
    Construction phase
    Pheromone update phase
    Optional daemon action phase
end
Return the best solution
```

3.1 Construction Phase

In this phase, artificial ants are created. An ant is assumed to be a sequence of states. For instance, if the objective is to find the shortest loop in a graph while visiting all nodes, four artificial ants are show in Fig. 3. This figure shows that an artificial ant is just a solution to this problem.

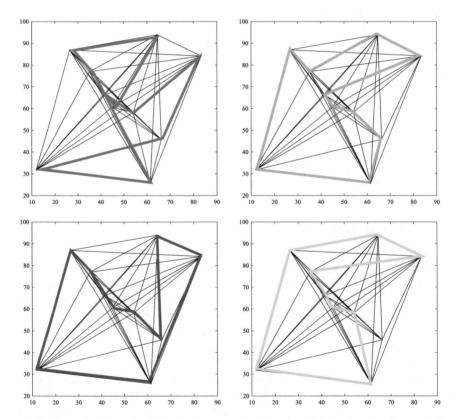

Fig. 3 Four artificial ants for the TSP problem with 10 cities, in which the shortest path to visit all cities should be found

The above example showed that artificial ants are created from a finite set of n available solutions [9]. Each ant can be seen as a set of values taken from the main finite set. In the construction phase, a component from the main set is selected and added to the artificial ant. The process is done based on the solution construction mechanism.

The above example is in fact the Traveling Salesman Problem (TSP) [19, 20], in which the shortest path between multiple "cities" should be found. To generate an artificial ants in each iteration of optimization, an ant can choose all the unvisited edges available from the current node. This selection process is done using the following probability:

$$p_{i,j} = \frac{(\tau_{i,j}^{\alpha})(\eta_{i,j}^{\beta})}{\sum (\tau_{i,j}^{\alpha})(\eta_{i,j}^{\beta})} \tag{1}$$

where $\tau_{i,j}$ shows the amount of pheromone of i, j edge, α defined the impact of pheromone, $\eta_{i,j}$ indicates the desirability of i, j edge, and β defines the impact of the desirability.

The desirability of an edge is defined by a weighting function. This function is mostly heuristic and assigns a value which shows how good an edge is. If the distance should be minimized, $\eta_{i,j}$ can be defined as follows:

$$\eta_{i,j} = \frac{1}{d_{i,j}} \tag{2}$$

where $d_{i,j}$ is the length of the edge i, j.

3.2 Pheromone Phase

This phase simulates the process of depositing and evaporation pheromone in nature. We use it to establish communication medium between ants [10]. In the example above, the amount of pheromone on i, j edge is calculated as follows:

$$\tau_{i,j} = (1 - \rho)\tau_{i,j} + \Delta\tau_{i,j} \tag{3}$$

where ρ defines the rate of pheromone evaporation and $\Delta\tau_{i,j}$ is the total amount of pheromone deposited on the i, j edge defined as follows:

$$\Delta\tau_{i,j} = \begin{cases} \frac{1}{L_k} & k\text{th ant travels on edge } i, j \\ 0 & \text{otherwise} \end{cases} \tag{4}$$

where L_k shows the length (cost value) of the path that kth ant travels.

A general equation for the pheromone phase in which the pheromone values are defined by all ants (with a tour in TSP for instance) is written as follows:

$$\tau_{i,j} = (1 - \rho)\tau_{i,j} + \sum_{k=1}^{m} \Delta\tau_{i,j}^{k} \tag{5}$$

where m shows the number of ants, ρ is the rate of pheromone evaporation, and $\Delta\tau_{i,j}^{k}$ indicates the amount of pheromone deposited by the kth ant on edge i, j.

In the above equations, the greater the value of ρ, the higher the evaporation rate. Also, to calculate the pheromone on each edge, the route of each ant should be considered.

3.3 Daemon Phase

The first two mechanism allows artificial ants to change their routes depending on the desirability of a node (closer distance in case of TSP) and pheromone level. However, the algorithm is at the risk of losing a very good route due to the consideration of probabilities in the mathematical model. To void loosing such good solutions, Darmon phase can be used. Different techniques can be used in this phase to bias the search. For instance, the best path in each iteration can be maintained or other local searches can be integrated to each artificial ant [11]. This step is very similar to elitism in evolutionary algorithms. Using such mechanism is not derived from the intelligence of ants. They can be seen as a centralized control unit that are used to assist ACO in solving problems.

The original version of ACO (Ant System) was developed using the above-mentioned mathematical model. However. there has been a lot of improvements afterwards [12]. Three of them are discussed as follows:

3.4 Max-Min Ant System

In this version of Ant System [13], there is a range for the pheromone concentration and the best ant is only allowed to lay pheromone. In the Max-Min Ant System algorithm, the equation for updating pheromone is defined as follows:

$$\tau_{i,j} = \left[(1 - \rho)\tau_{i,j} + \Delta\tau_{i,j}^{best}\right]_{\tau_{lb}}^{\tau_{ub}} \tag{6}$$

where $\Delta\tau_{i,j}^{best}$ is the total amount of pheromone deposited on the i, j edge by the best ant, τ_{ub} is the upper bound of pheromone, τ_{lb} shows the lower bounds pheromone, and $[x]_{\tau_{lb}}^{\tau_{ub}}$ is defined as follows:

$$[x]_{\tau_{lb}}^{\tau_{ub}} = \begin{cases} ub & x > ub \\ lb & x < lb \\ x & \text{otherwise} \end{cases} \tag{7}$$

This component is similar to velocity clamping in the PSO algorithm [14].

The summation of pheromone deposited by the best ant is calculated as follow (note that the best ant can be selected from the set of best ants in the current iteration or the best ant obtained so far):

$$\Delta \tau_{i,j}^{best} = \begin{cases} \frac{1}{L_{best}} & \text{the best ant travels on edge } i, j \\ 0 & \text{otherwise} \end{cases} \tag{8}$$

where L_{best} shows the distance of the best path.

3.5 Ant Colony System

In the Ant Colony System [7], the focus was mostly on the Daemon phase, in which a mathematical model was employed to perform local pheromone update. In this approach, the following equation requires the ants to decrease the pheromone level so that the subsequent ants gravitate toward other routes:

$$\tau_{i,j} = (1 - \varphi)\tau_{i,j} + \varphi\tau_0 \tag{9}$$

where τ_0 shows that initial pheromone level and φ is a constant to reduce the pheromone level.

The process of decaying pheromone [15] by each ant promotes exploration of the search space since it reduces the probability of choosing the same path by subsequent ants. Also, evaporation of pheromone prevents premature and rapid convergence.

3.6 Continuous Ant Colony

The easiest way to use ACO for solving continuous problems is to split the variables' domain into a set of intervals [16]. This technique basically discretized the continuous domain which results in low accuracy specially for problems with wide-domain variables. However, a combinatorial algorithm can be easily used without modification. For instance, the domain of variable x in the interval of [0, 10] can be discretized into $s = \{0, 1, 2, .., 10\}$. The accuracy of this technique can be improved with increasing the number of values in the set. Several continuous versions of ACO have been developed in the literature to solve continuous problems without domain modification. Interested readers are referred to [17, 18].

4   Application of ACO in AUV Path Planning

In this section, the ACO algorithm is employed to find the shortest path for an AUV to achieve an objective. The main objective of the AUV is to reach a certain number of way-points underwater while minimizing the battery usage. To minimize the amount of electricity used by the vehicle, the shortest path should be found while visiting all the way-points once. The way-points are generated randomly in each experiment to benchmark the accuracy and robustness of the solver. The number of way-points are changed from 5 to 200 with the step size of 5, so a total of 40 case studies are used in this subsection. Thirty ants and 50 iterations are used to obtain the results in the tables.

The Results of ACO when the number of way-points are equal to 5, 10, 15, ..., 195, 200 are given in Tables 1 and 2. These two tables show the average, standard deviation, and median of the distances between the way-points in the shortest path

Table 1 Results of ACO when the number of way-points is equal to 5, 10, 15, ..., 95, 100. The average, standard deviation, and median of the distances of between the way-points in the shortest path obtained by ACO after 10 independent runs are given. The average run time over these 10 runs when using one- or eight-core CPU are given as well

Way-points#	Distance (m)			Time (s)	
	Mean	Std	Median	1 core	8 cores
5	215.48	40.53	219.54	0.13	0.02
10	300.60	29.12	291.94	0.24	0.03
15	347.06	26.50	347.04	0.38	0.05
20	421.05	35.75	421.31	0.52	0.06
25	478.02	25.17	480.04	0.67	0.08
30	538.36	52.38	544.50	0.77	0.10
35	599.42	34.36	612.81	0.90	0.11
40	657.76	41.03	663.52	1.09	0.14
45	706.25	36.97	714.34	1.33	0.17
50	751.17	53.14	757.23	1.38	0.17
55	788.42	33.73	788.93	1.47	0.18
60	835.49	41.51	845.06	1.62	0.20
65	897.53	32.91	903.32	1.76	0.22
70	944.38	16.28	945.02	1.91	0.24
75	964.72	24.11	959.72	2.06	0.26
80	1017.20	51.04	1009.66	2.29	0.29
85	1070.34	34.43	1067.40	2.37	0.30
90	1114.62	50.72	1120.04	2.61	0.33
95	1166.05	36.02	1182.78	2.71	0.34
100	1221.85	38.93	1224.31	2.85	0.36

obtained by ACO after 10 independent runs. The average run time over these 10 runs when using one- or eight-core CPUs are given as well.

Table 1 shows that the ACO performed well on all of the case studies. The length of the shortest path increases proportional to the number of way-points, which is due to the more number of places that the AUV needs to visit. The CPU time required to find the shortest path is increased as well. The runtime of different phases of ACO and calculating the objective function highly depends on the number of way-points. the run time does go above one second after considering 40 way-points. One way to reduce this to be able to have a more practical use of the ACO algorithm is parallel processing. As an example, eight nodes are used to do the same experiments and the results show that the run time can be significantly decreased. At the best case the run-time will be T/n where T is the runtime with one node and n is the number of nodes.

Table 2 Results of ACO when the number of way-points are equal to 100, 105, 110, ..., 195, 200. Average, standard deviation, and median of the distances of between the cities in the shortest path obtained by ACO after 10 independent runs are given. The average run time over these 10 runs when using one- or eight-core CPU are given as well

Way-points#	Distance (m)			Time (s)	
	Mean	Std	Median	1 core	8 cores
100	1234.27	47.11	1221.31	2.84	0.35
105	1268.12	53.02	1279.53	3.04	0.38
110	1293.54	33.06	1302.21	3.24	0.40
115	1351.58	89.81	1373.99	3.38	0.42
120	1380.83	56.74	1378.97	3.75	0.47
125	1419.62	61.83	1405.56	3.93	0.49
130	1453.15	52.02	1464.68	4.11	0.51
135	1499.80	78.28	1504.41	4.14	0.52
140	1540.98	50.22	1538.96	4.70	0.59
145	1575.39	48.48	1566.75	4.43	0.55
150	1590.36	39.57	1585.83	4.61	0.58
155	1626.32	54.68	1630.07	4.78	0.60
160	1693.89	47.15	1693.80	4.97	0.62
165	1684.74	88.52	1677.57	5.13	0.64
170	1751.73	62.09	1752.78	5.38	0.67
175	1773.97	54.67	1779.71	5.76	0.72
180	1851.22	110.91	1866.75	5.83	0.73
185	1870.89	45.41	1866.55	6.06	0.76
190	1853.45	71.67	1845.45	6.26	0.78
195	1941.50	55.50	1948.24	6.94	0.87
200	1971.69	41.18	1960.24	7.13	0.89

To see the shape of the shortest path obtained by ACO on the case studies in Table 1, Fig. 4 is provided. This figure shows that the best path obtained by the ACO algorithm is highly near optimal. The stochastic nature of ACO leads to different results in each run. With running this algorithm several times and fine tuning this algorithm, the best performance of ACO on each of the case studies can be achieved. In this experiment, 30 ants and 50 iterations are used. These numbers lead the shortest path when the number of way-points is small. For a large number of way-points, however, the path obtained is near optimal. This is evident in the last subplot of Fig. 4. The path obtained is reasonable, but it is not the shortest. However, it should be noted that this high-quality solution has been obtained in a very short period of time, which is the main advantage of heuristics and meta-heuristics as compared to deterministic algorithms.

The convergence curves are also visualized in Fig. 4. Such figures show how the first best initial path is improved over the course of iterations. Except the first curve with a very sharp decline, other curves are steady. This first case study is very simple due to the use of five way-points only. For other case studies, however, the algorithm does not quickly converge towards the global optimum. The algorithm slowly searches the search space, finds promising regions, and exploits them.

The rest of the results are presented in Table 2 and Fig. 5. The results of this table are worse than those in Table 1. This is again due to the larger number of way-points. The case studies in this table are very challenging. This causes the average speed to reaches up to seven seconds, which can be reduced significantly using parallel processing. The best path obtained by ACO in Fig. 5 show that the best paths are relatively optimal. The distance between most of the near way-points are optimal, but these case studies require a larger number of ants and iterations.

The convergence curves in Fig. 5 are different from those in Fig. 4, in which that rate of convergence significantly changes after in the first half of the iterations. This is again due to the difficulty of the case studies. The algorithm starts improving the quality of the first initial population. As the algorithm moves towards the final iterations, however, the algorithm needs to tune the sub-optimal paths obtained so far. In case of stagnation in local solutions and high pheromone concentration, it is difficult for the ACO algorithm to find a different path. Introducing random changes (*e.g.* mutation) might help, yet they are outside the scope of this chapter. The accuracy of the solution obtained by ACO and run-time are acceptable.

To see the impact of evaporation parameter on the performance of ACO, the case study with 100 way-points is solved with changing this parameter. After every change in the evaporation rate, the algorithm is run 10 times. ACO is equipped with 50 ants and 200 iterations. As results, average, standard deviation, median, minimum, and maximum of the length of the shortest path obtained at the end of each iteration after every run are presented in Table 3 and Fig. 6.

It can be seen that the accuracy of the results is reduced proportionally to the value of the evaporation constant. Increasing the evaporation constant leads to more random decisions of artificial ants, which is essential to solve large-scale problems with a large number of local solutions. However, the bar charts in Fig. 6 show that

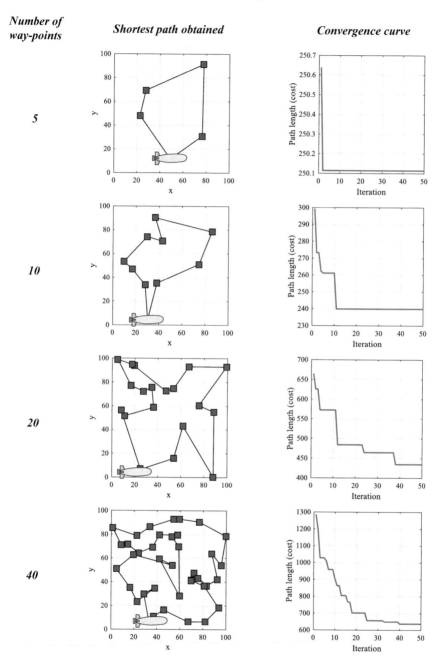

Fig. 4 The shortest path and convergence curve obtained by ACO on small-scale case studies

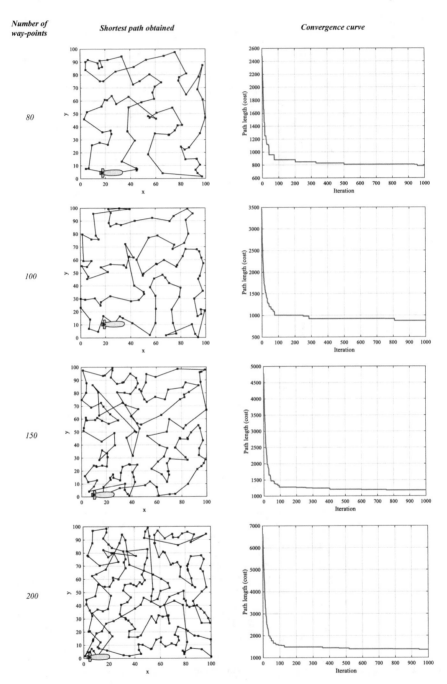

Fig. 5 The shortest path and convergence curve obtained by ACO on large-scale case studies

Table 3 Average, standard deviation, median, best, and worst of the best solution obtained in 10 runs when changing the evaporation constant

Evaporation constant	Length of the best path				
	Mean	Std	Median	Best	Worst
$\varphi = 0.05$	1473.85	20.48	1476.86	1446.90	1499.91
$\varphi = 0.1$	1425.49	32.26	1412.75	1391.22	1478.53
$\varphi = 0.2$	1338.00	52.31	1350.88	1238.46	1403.63
$\varphi = 0.4$	1256.03	36.54	1248.02	1222.66	1345.75
$\varphi = 0.6$	1219.11	29.44	1217.82	1169.96	1262.38
$\varphi = 0.8$	1251.90	29.95	1252.77	1210.71	1298.02

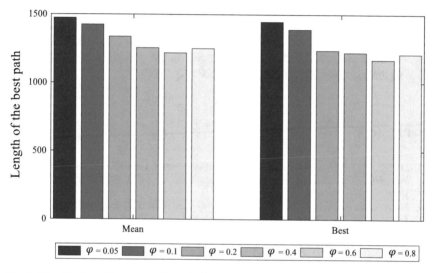

Fig. 6 Visualization of the mean and best obtained path over 10 runs when changing the evaporation constant (φ)

increasing this parameter to more than 0.6 will degrade the quality of the solution. This is due to tapering off the impact of the main search mechanisms of ACO when evaporation is very high.

Overall, the results show the merits of the ACO algorithm in finding the shortest path. This algorithm has been widely used in similar problems in the literature. As per the result of this chapter, we recommend it to be used in path planning operators of autonomous vehicles when solving challenging navigation-related problems where conventional search methods are not practical.

5 Conclusion

This chapter discusses the main mechanisms of the ACO Algorithm as one of the most well-regarded combinatorial optimization algorithms. It was discussed that this algorithm simulates the social intelligence of ants in finding the shortest path from a nest to a food source. Since ACO is a stochastic algorithm, it benefits from several random components and probabilistic models. Such mechanisms prevent this algorithm from stagnation in local solutions. After the presentation and brief literature review of ACO, it was applied to 20 case studies. The case studies were a real application of TSP, in which an AUV was required to find the shortest path for traveling through multiple way-points. Up to 200 way-points were considered to see how efficient ACO is on large-scale version of this problem. The ACO algorithm performed well and found the shortest path in a short period of time. This is opposed to deterministic techniques (*e.g.* brute-force search) where the runtime increases exponentially proportional to the number of way-points to visit. The chapter also investigated the impact of the evaporation rate on the performance of AVO when solving the most difficult case study with 200 way-points.

References

1. Dorigo, M., & Di Caro, G. (1999). Ant colony optimization: A new meta-heuristic. In *Proceedings of the 1999 Congress on Evolutionary Computation, 1999 CEC 99* (Vol. 2, pp. 1470–1477). IEEE.
2. Dorigo, M., & Birattari, M. (2011). Ant colony optimization. In *Encyclopedia of Machine Learning* (pp. 36–39). Springer, Boston, MA.
3. Grass, P. P. (1959). La reconstruction du nid et les coordinations interindividuelles chezBellicositermes natalensis etCubitermes sp. la thorie de la stigmergie: Essai d'interprtation du comportement des termites constructeurs. *Insectes sociaux, 6*(1), 41–80.
4. Dorigo, M., Bonabeau, E., & Theraulaz, G. (2000). Ant algorithms and stigmergy. *Future Generation Computer Systems, 16*(8), 851–871.
5. Dorigo, M., Maniezzo, V., & Colorni, A. (1996). Ant system: Optimization by a colony of cooperating agents. *IEEE Transactions on Systems, Man, and Cybernetics, Part B (Cybernetics), 26*(1), 29–41.
6. Stützle, T., & Hoos, H. H. (1996). Improving the ant system: A detailed report on the MAXMIN Ant System. FG Intellektik, FB Informatik, TU Darmstadt, Germany, Tech. Rep. AIDA9612.
7. Dorigo, M., & Gambardella, L. M. (1997). Ant colony system: A cooperative learning approach to the traveling salesman problem. *IEEE Transactions on Evolutionary Computation, 1*(1), 53–66.
8. Papadimitriou, C. H., & Steiglitz, K. (1998). Combinatorial optimization: Algorithms and complexity. *Courier Corporation.*
9. Dorigo, M., & Stützle, T. (2003). The ant colony optimization metaheuristic: Algorithms, applications, and advances. In *Handbook of metaheuristics* (pp. 250–285). Springer, Boston, MA.
10. Stützle, T. (2009, April). Ant colony optimization. In *International Conference on Evolutionary Multi-Criterion Optimization* (pp. 2). Springer, Berlin, Heidelberg.
11. Stützle, T., Lpez-Ibnez, M., Pellegrini, P., Maur, M., De Oca, M. M., Birattari, M., & Dorigo, M. (2011). Parameter adaptation in ant colony optimization. In *Autonomous Search* (pp. 191–215). Springer, Berlin, Heidelberg.

12. Randall, M., & Lewis, A. (2002). A parallel implementation of ant colony optimization. *Journal of Parallel and Distributed Computing, 62*(9), 1421–1432.
13. Stützle, T., & Hoos, H. H. (2000). MAXMIN ant system. *Future Generation Computer Systems, 16*(8), 889–914.
14. Shahzad, F., Baig, A. R., Masood, S., Kamran, M., & Naveed, N. (2009). Opposition-based particle swarm optimization with velocity clamping (OVCPSO). In *Advances in Computational Intelligence* (pp. 339–348). Springer, Berlin, Heidelberg.
15. Sharvani, G. S., Ananth, A. G., & Rangaswamy, T. M. (2012). Analysis of different pheromone decay techniques for ACO based routing in ad hoc wireless networks. *International Journal of Computer Applications, 56*(2),
16. Socha, K. (2004, September). ACO for continuous and mixed-variable optimization. In *International Workshop on Ant Colony Optimization and Swarm Intelligence* (pp. 25–36). Springer, Berlin, Heidelberg.
17. Socha, K., & Dorigo, M. (2008). Ant colony optimization for continuous domains. *European Journal of Operational Research, 185*(3), 1155–1173.
18. Blum, C. (2005). Ant colony optimization: Introduction and recent trends. *Physics of Life Reviews, 2*(4), 353–373.
19. Hoffman, K. L., Padberg, M., & Rinaldi, G. (2013). Traveling salesman problem. In *Encyclopedia of Operations Research and Management Science* (pp. 1573–1578). Springer US.
20. Reinelt, G. (1991). TSPLIBA traveling salesman problem library. *ORSA Journal on Computing, 3*(4), 376–384.

Ant Lion Optimizer: Theory, Literature Review, and Application in Multi-layer Perceptron Neural Networks

Ali Asghar Heidari, Hossam Faris, Seyedali Mirjalili, Ibrahim Aljarah and Majdi Mafarja

Abstract This chapter proposes an efficient hybrid training technique (ALOMLP) based on the Ant Lion Optimizer (ALO) to be utilized in dealing with Multi-Layer Perceptrons (MLPs) neural networks. ALO is a well-regarded swarm-based meta-heuristic inspired by the intelligent hunting tricks of antlions in nature. In this chapter, the theoretical backgrounds of ALO are explained in details first. Then, a comprehensive literature review is provided based on recent well-established works from 2015 to 2018. In addition, a convenient encoding scheme is presented and the objective formula is defined, mathematically. The proposed training model based on ALO algorithm is substantiated on sixteen standard datasets. The efficiency of ALO is compared with differential evolution (DE), genetic algorithm (GA), particle swarm optimization (PSO), and population-based incremental learning (PBIL) in terms of best, worst, average, and median accuracies. Furthermore, the convergence propensities are monitored and analyzed for all competitors. The experiments show that

A. A. Heidari
School of Surveying and Geospatial Engineering, University of Tehran, Tehran, Iran
e-mail: as_heidari@ut.ac.ir

H. Faris · I. Aljarah
King Abdullah II School for Information Technology, The University of Jordan, Amman, Jordan
e-mail: hossam.faris@ju.edu.jo

I. Aljarah
e-mail: i.aljarah@ju.edu.jo

S. Mirjalili (✉)
Institute of Integrated and Intelligent Systems, Griffith University, Nathan, Brisbane, QLD 4111, Australia
e-mail: seyedali.mirjalili@griffithuni.edu.au

M. Mafarja
Faculty of Engineering and Technology, Department of Computer Science, Birzeit University, PoBox 14 Birzeit, Palestine
e-mail: mmafarja@birzeit.edu

© Springer Nature Switzerland AG 2020
S. Mirjalili et al. (eds.), *Nature-Inspired Optimizers*, Studies in Computational Intelligence 811, https://doi.org/10.1007/978-3-030-12127-3_3

23

the ALOMLP outperforms GA, PBIL, DE, and PSO in classifying the majority of datasets and provides improved accuracy results and convergence rates.

1 Introduction

The brain of physicists can realize the puzzle of various concurrent and complicated events in nature. The reason of why such intelligence can be raised among human being is that the evolution has slowly but surely enhanced the architecture of our brains as an immensely parallel network of neurons [80]. This was a revolutionary fact that originally inspired McCulloch and Pitts in 1943 to build a mathematical paradigm for simulating the biological nervous system of the brain, which also encouraged other researchers to investigate on artificial neural network (ANN) [36, 66].

Artificial neural networks (ANNs) [66] are well-known techniques utilized in learning, approximating, and investigating various classes of complex problems [7, 14, 20, 28, 33, 45, 91]. It is validated that ANNs can perform generally well in dealing with machine perception scenarios, where it is hard to individually interpret the main features of the problem[28]. Hence, ANNs are accepted as powerful learning methods for divulging satisfactory results on many pattern recognition, clustering, classification, and regression problems [7, 12, 13, 18, 22, 59, 60]. Single layer perceptron (SLP) networks are computational models that contain only two input and output layers, which this is the simplest form of ANNs [52]. It has been proved that the SLPs are not capable of efficiently detecting and tackling the nonlinearly separable patterns [73]. In order to mitigate the problems of SLPs, researchers developed a special from of ANNs, which is called multilayer perceptron (MLP). The MLP networks often utilize several hidden layers to avoid the drawbacks of SLPs. Consequently, MLPs are known as the most employed class of ANNs in literature [19]. The main advantages of MLP are the high learning potential, robustness to noise, nonlinearity, parallelism, fault tolerance, and high capabilities in generalizing tasks [31]. Based on Kolmogorov's theorem, MLP with a single hidden layer can estimate any continuous function [21]. However, the performance of ANNs are very dependent on the learning technique utilized to train the weighting vectors [73].

The supervised MLP trainers are often originated from two official classes: gradient-based and stochastic-based algorithms [73]. While gradient-based trainers have a good performance in local search [102], meta-heuristic-based trainers (meta-trainers) can reveal a high efficiency in avoiding local optima (LO) [1, 7, 33]. Randomness of search can be considered as the basic advantage of meta-trainers over gradient-based trainers [11, 29, 44, 63]. Therefore, they can carry out global search usually better than gradient-based trainers [40]. They are straightforward and flexible [49]. Furthermore, when the prior knowledge about the task is not presented, meta-trainers are often the best choices [58, 73]. Gradient-based techniques can only address the solutions of continuous and differential tasks, while the meta-heuristic optimizers can also tackle nonlinear and non-differentiable functions. However, they still need more computational time [31, 33, 47, 48].

Enhancing the structure and weights of a MLP simultaneously by the trainer algorithm faces a large-scale task [73]. Based on previous works, evolutionary algorithms (EA) and swarm-based metaheuristic algorithms (SMHAs) have been widely integrated with machine learning algorithms such as MLP [2, 6, 37, 38, 64, 83]. The SMHAs inspire the team-based and self-organized behaviors of animals who live as a swarm such as birds, grasshoppers, whales, and wolves [30, 46, 49, 70]. These algorithms are used to enhance the weights, parameters, and structure of MLP. A remarkable fact about meta-heuristic algorithms (MHAs) is that one optimizer can often outperform other MHAs only in tackling a limited set of problems, that is, "there is no such universal optimizer which may solve all class of problems" [95], which is a significant consequence of No Free Lunch (NFL) theorem [95]. The NFL theory confirms that "the average performance of any pair of algorithms across all possible problems is identical" [95]. Hence, a result of NFL is as follows:"A general purpose universal optimization strategy is impossible, and the only way, one strategy can outperform another if it is specialized to the structure of the specific problem under consideration" [50]. However, many researchers are still motivated to launch new MHAs for handling the ANNs. The key reason is that NFL holds only for some categories of problems. In NFL, a problem should be closed under permutation (cup) [79]. Therefore, a new MHA can outperform other MHAs for those problems that are not cup [9, 53, 73]. Note that most of the real-world tasks in science are not cup [10, 32, 45, 53, 73].

Several works utilized the genetic algorithm (GA) based to evolve their proposed ANNs [3, 81, 82, 84, 93]. Differential evolution (DE) [65] and evolution strategy (ES) [43] are also utilized as base trainers in several papers [54, 85, 92, 94]. The results affirm the satisfactory results of these hybrid MLP structures. Particle swarm optimizer (PSO), artificial bee colony (ABC), and ant colony optimization (ACO) also show a competitive efficacy in dealing with optimizing MLPs [15, 16, 56, 73, 86, 100]. Lightning search algorithm (LSA) [31], gray wolf optimizer (GWO) [68], social spider optimizer (SSO) [71], whale optimizer (WOA) [7], Monarch Butterfly Optimization (MBO) [35], Biogeography-based Optimizer (BBO) [8], and Moth-flame Optimizer (MFO) [34] are also some of recent trainers in literature.

Searching for global results of hybrid MLP networks is still an open question [31, 33]. From NFL theorem [95], a new superior SMHA can still be designed to be integrated with MLP networks. Motivated by this these facts, this book chapter is devoted to design a novel ALO-based training method for MLP networks. ALO is a successful SMHA invented by Mirjalili in 2015 [67]. Up to 2018, ALO has revealed very encouraging outcomes in dealing with multifaceted and benchmark problems [4].

2 Ant Lion Optimizer

Ant lion optimizer (ALO) is a new successful population-based MHA which tries to mimic the idealized hunting activities of antlions in nature [67]. The name of this insect originates from its exceptional hunting tactics in addition to its beloved

quarry. They dig a cone-shaped fosse in sand based on some circular motions with
their powerful jaw. After building a cone-shaped trap, these sit-and-wait hunters will
hide themselves at the bottom of cone and wait for often ants to be trip over their trap
with sharp edges. After detecting that an insect cannot escape the cone-shaped pit,
they attack to kill the prey. As insects often attempt to survive from the faced traps in
nature, they will perform some abrupt movements to get away from predator. At this
stage, antlions intelligently start to toss sands just before the edges of the channel to
slide the frightened victim into the deepest point of the hole. When a quarry cannot
increase his safe distance to the jaw, it will be dragged under the surface and killed.
Extra thought-provoking behavior performed by these hunters is the proportion of
the dimensions of the pit with two factors: degree of malnutrition and phase of the
moon. Referring to the above behaviors and tricks, the ALO has two classes of search
agents, namely ant and antlions. The best search agents are selected as antlions which
never change their positions except the situation they substitute a particular ant. Ant
agents can carry out random walks in the solution space and may be captured by
an antlion if be trapped in a pit. The ants location can be attained using the rule in
Eq. (1):

$$Ant_i^t = \frac{R_A^t + R_E^t}{2} \tag{1}$$

where R_A^t is the random walk nearby the target antlion nominated by the roulette
wheel at t-th iteration, R_E^t is the location of randomly walking ant indexed i, named
Ant_i, nearby the elite antlion indexed E in the swarm of ants. Fitness values of antlions
are utilized by the roulette wheel mechanism to select an antlion with index A to
execute the ant random walk nearby it whereas the elite antlion is recognized as the
best antlion with index E. The random walking behavior of an Ant_i^t in the vicinity of
an assumed $Antlion_j^t$ can be expressed by Eq. (2):

$$R_j^t = \frac{(X_i - a_i) \times (d_i - c_i^t))}{(b_i^t - a_i)} + c_i \tag{2}$$

where R_j^t is the location of ant i after carrying out random walk nearby antlion j
during iteration t, a_i is the minimum step of random walk X_i^t in i-th dimension, and
b_i is the maximum step of random walk X_i^t in i-th dimension, X_i is described in Eq.
(3), c, d are the inferior and superior restraints of the random walk, respectively.
Equation (2) can be directly utilized to realize R_A^t and R_E^t of Eq. (1).

$$X(t) = [0, CS(2\kappa(t_1) - 1); CS(2\kappa(t_2) - 1); \cdots ; CS(2\kappa(t_T) - 1)] \tag{3}$$

where CS shows the cumulative sum and generates the accumulation of successive
random jumps creating a random walk pending time t, and $\kappa(t)$ is a randomized
function considered as in Eq. (4):

$$\kappa(t) = \begin{cases} 1 & r > 0.5 \\ 0 & r \le 0.5 \end{cases} \tag{4}$$

where r is a random number inside $(0, 1)$, c, d parameters are adapted according to Eqs. (5) and (6) to control the variety of the random walk nearby the assumed antlion.

$$c_i^t = \begin{cases} \frac{lb}{\delta} + X_{Antlion_j^t} & r' < 0.5 \\ \frac{-lb}{\delta} + X_{Antlion_j^t} & otherwise, \end{cases} \tag{5}$$

$$c_i^t = \begin{cases} \frac{ub}{\delta} + X_{Antlion_j^t} & q > 0.5 \\ \frac{-ub}{\delta} + X_{Antlion_j^t} & otherwise, \end{cases} \tag{6}$$

where r' and q are random numbers in $(0,1)$, lb and ub are the inferior and superior bounds for dimension i, respectively, δ is a factor that can monitor the diversification/intensification ratio and is formulated as in Eq. (7):

$$\delta = 10^w \frac{t}{T_{max}} \tag{7}$$

where T_{max} shows the maximum number of iterations, w denotes a constant represented based on the present iteration ($w = 2$ when $t > 0.1T$, $w = 3$ when $t > 0.5T$, $w = 4$ when $t > 0.75T$, $w = 5$ when $t > 0.9T$, and $w = 6$ when $t > 0.95T$). The constant w is capable of adjusting the accuracy of intensification. By increasing the iteration count, the radius of random walk will decrease, which this fact can guarantee convergence of this method. Lastly, the selection process is performed where an antlion is substituted by an ant if the ant be a better solution. The pseudo-code of ALO is represented in Algorithm 1.

Algorithm 1 Pseudo-code of the ALO algorithm

Input: Total number of ants and antlions, and number of iterations (T_{max}).
Output: The elitist antlion and the corresponding fitness value.
Initialize the random positions of all agents $x_i (i = 1, 2, \ldots, n)$ inside ub and lb bounds.
Calculate the fitness of population.
Select the Elite antlion E.
while (end condition is not met) **do**
    for (each Ant_i) **do**
        Select an antlion A based on roulette wheel.
        Run random walk of Ant_i nearby Antlion A by Eq. (3.2);R_A^t
        Run random walk of Ant_i nearby Antlion A by Eq. (3.2);R_E^t
        Update the location of Ant_i using Eq. (3.1).
    Update the fitness values of all agents
    Merge all ants and sort them based on the fitness metric and select the fittest n agents as the new antlions and the worst n agents as the ants.
    Update the elite ant if an antlion is better than the elite agent.

---

3 Literature Review

In this section, previous works on ALO and their main results are reviewed in details. Zawbaa et al. [27] proposed a binary ALO for feature selection tasks. The results show that binary ALO can significantly outperform the conventional PSO and GA. Zawbaa et al. [101] also improved the efficiency of ALO in treating feature selection tasks by proposing a new chaos-based ALO (CALO). Gupta et al. [41] developed a new ALO-based toolkit in LabVIEW^{TM}. Yamany et al. [97] used ALO to optimize the connection vectors of MLP. They used only four datasets and the results expose that ALO can outperform ACO, PSO, and GA. Rajan and Malakar [76] proposed a weighted elitism-based ALO (MALO) to treat Optimal Reactive Power Dispatch (ORPD) tasks in power systems and the results affirms the robustness and consistency of the MALO.

Kamboj et al. [55] also employed ALO to efficiently solve a non-convex ORPD problem. Ali et al. [5] utilized basic ALO to manage the best allocation and sizing of distributed generation (DG). The comparative results revealed the dominance of ALO over other metaheuristics such as Artificial Bee Colony (ABC), Firefly Algorithm (FA), Cuckoo Search (CS), Ant Colony Optimization (ACO), GA and PSO. The basic ALO has been also utilized to manage the integrated planning and scheduling tasks [75]. Raju et al. [77] developed an ALO-based approach for simultaneous optimization of controllers. Dubey et al. [24] applied basic ALO to hydro-thermal-wind scheduling (HTPGS) tasks. The results show the capabilities and advantages of ALO in realizing better-quality solutions while satisfying a lot of of real-world restrictions of hydro, thermal, and wind generation.

Mirjalili et al. [69] proposed the multi-objective variant of ALO (MOALO) and results reveal the high efficacy of this method in realizing the solutions of real-world engineering problems. Talatahari [87] used ALO to deal with optimal design of skeletal constructions and results show superior efficacy of ALO compared to several classical algorithms. Kaushal and Singh [57] investigated the performance of ALO in dealing with optimal allocation of stocks in portfolio and the results show that ALO can outperform GA, efficiently. Saxena and Kothari [78] investigated the efficacy of ALO on antenna array synthesis problem and their results show excellent performance of ALO compared to previous methods such as PSO, ACO, and biogeography based optimization (BBO). Yao and Wang [98] proposed a new adaptive ALO (DAALO) for planning of unmanned aerial vehicles (UAV) based on the levy flight concept and revising the extent of traps. The results show high performance and robustness of the new DAALO. Hu et al. [51] proposed an efficient approach (ALODM) for big data analytics by integrating ALO with differential mutation process. In ALODM, differential mutation operations and greedy scheme has enhanced the diversity of the antlions.

Dinkar and Deep [23] proposed an opposition-based Laplacian ALO (OB-L-ALO) for numerical optimization and results show the improved performance of OB-L-ALO compared to conventional ALO. Wu et al. [96] proposed an improved ALO

(IALO) to realize the parameter of photovoltaic cell models. The IALO algorithm can arrange the initial antlions using chaos-triggered signals to deepen the uniformity and ergodicity of swarm. In addition, they inspired the motion rule of PSO to update the positions of antlions to improve the intensification and diversification tendencies of ALO. Results show that IALO can significantly outperform ALO, PSO and Bat Algorithm (BA). Cao et al. [17] developed a multi-objective ALO-based optimizer integrated with SVMs, which is called SVM/ALO/GH, to attain the set of Pareto optimal parameters in gear hobbing and the results show that ALO-based model can outperform DE-based method. Nair et al. [72] utilized ALO for adaptive optimization of infinite impulse response (IIR) filters and results show that the proposed ALO-based method can efficiently outperform the alternative variant which works based on gravitational search algorithm (GSA). Trivedi et al. [90] applied the conventional ALO to optimal power flow (OPF) problem. The simulations show that ALO can provide better results than PSO and FA in dealing with OPF problem.

In 2018, Elaziz et al. [26] integrated conventional ALO with adaptive neuro-fuzzy inference system (ANFIS) to generate a Quantitative structure-activity relationship (QSAR) Model for HCVNS5B inhibitors. The results, which are published in scientific reports nature, showed that the satisfactory efficacy of the proposed ALO-based QSAR model with regard to different metrics. Tian et al. [89] proposed a modified ALO for parameter identification of Hydraulic Turbine Governing Systems. Li et al [61] utilized a two-stage MOALO-based optimizer for solving optimal DGs planning problem. The results on the PG&E 69-bus distribution system reveal that the MOALO has a better efficacy compared to other methods. Tharwat and Hassanien [88] proposed a chaotic ALO (CALO) for optimal tuning of support vector machines (SVMs) and results show that CALOSVM classifier can vividly outperform GASVM and PSOSVM variants.

Gandomi and Kashani [39] applied several recent optimizers including ALO to construction rate optimization of shallow foundation and results show that ALO can provide very competitive results. Dubey et al. [25] also applied several new optimizers including ALO to wind integrated multi-objective power dispatch (MOOD) problem. Results show that ALO is one of three best methods in tackling MOOD problem. Yogarajan and Revathi [99] proposed a cluster-based data gathering approach based on ALO to be utilized in Wireless Sensor Networks (WSNs). Oliva et al. [74] proposed an efficient ALO-based image segmentation technique. Hamouda et al. [42] applied ALO to kidney exchanges and they also provided open source software. The results expose that ALO-based technique can achieve to similar kidney exchange solutions compared to the deterministic-based algorithms such as integer programming. Moreover, ALO can provide superior results compared to other stochastic-based algorithms such as GA based on usage of computational resources and the quality of obtained exchanges.

4   Perceptron Neural Networks

Feedforward neural networks (FFNNs) are known as a well-regarded class of ANN-based neural models which are capable of realizing and approximating complex models based on their next-level, parallel, layered structure [73]. The fundamental processing elements of FFNNs are a series of neurons. These neurons are disseminated over a number of fully-linked loaded layers. MLP is one of the widespread examples of FFNNs. In MLP, the initial processing elements are prearranged according to a one-directional manner. In these networks, evolution of information happens based on the communications among three types of matching layers: input, hidden, and output layers. Figure 1 shows a MLP network that has a single hidden layer. The networks between these layers are associated with some weighting values varied inside [-1, 1]. Two functions can be carried out on every node of MLP, which are called summation and activation functions. The product of input values, weight values, and bias values can be attained based on the summation function described in Eq. (8).

$$S_j = \sum_{i=1}^{n} \omega_{ij} I_i + \beta_j \tag{8}$$

where n denotes the total number of inputs, I_i is the input variable i, β_j is a bias value, and w_{ij} reveals the connection weight.

In next step, an activation function is activated based on the outcome of the Eq. (9). Various activation approaches can be utilized in the MLP, which, according to literature, the most utilized one is S-shaped sigmoid function [7]. This function can be calculated based on Eq. (9).

$$f_j(x) = \frac{1}{1 + e^{-S_j}} \tag{9}$$

Fig. 1 Structure of MLP neural network

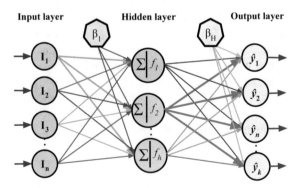

Therefore, the final output of the neuron j is attained using Eq. (10):

$$y_i = f_j(\sum_{i=1}^{n} \omega_{ij} I_i + \beta_j)$$ (10)

After building the final structure of ANN, the learning process is instigated to fine-tune and evolve the weighting vectors of network. These weighting vectors should be updated to approximate the results and optimize the total error of the network. Learning (training) step of the ANN is a computationally challenging process which has a significant impact on the efficacies and capabilities of the MLP in treating different problems.

5 ALO for Training MLPs

In this section, the structure and steps of proposed ALO-based MLP-trainer (ALOMLP) is developed and explained in detail. To integrate ALO with MLP, two technical points should be determined: how to encode the solutions and how to define the fitness function. In ALOMLP, all ants can be encoded as one-dimensional vectors consist of random numbers in $[-1, 1]$. Each ant will represent a candidate ANN. Figure 2 shows the encoding of ants in ALOMLP. As in Fig. 2, the encoded ant includes sets of connection weights and bias terms. The length of this vector is calculated based on the number of weights and biases in the MLP. Another observation is the generation of fitness function. To find the fitness of ants, all ants are sent to the MLP as the connection weights. The MLP evaluates those ants based on the utilized training dataset. Then, the MLP will reveal the fitness of all ants. The used fitness function in ALOMLP is the Mean Squared Error (MSE). For the training of samples, the MSE is obtained based on the variance of the real and predicted outcomes by the resulted ants (MLPs). The objective is to optimize the values of the MSE metric to some extent. The MSE metric is calculated by Eq. (11):

$$MSE = \frac{1}{n} \sum_{i=1}^{n} (z_i - \hat{z}_i)^2$$ (11)

where z is the actual value, $\hat{z}$ denotes the predicted value, and n is the number of instances.

The steps of ALOMLP trainer can be summarized by the following phases:

1. Initialization: the ALOMLP initiate random ants.
2. Mapping of ants: the elements of the ants are mapped to some weight and bias vectors of a candidate MLP.

A solution vector created by ALO

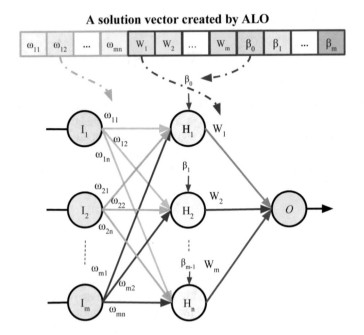

Fig. 2 Solution structure in ALOMLP

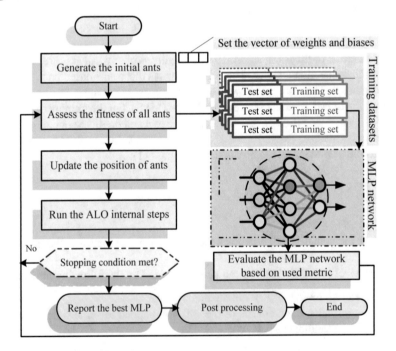

Fig. 3 MLP integrated with ALO

3. Fitness measure: the goodness of the MLPs is assessed according to the MSE for all samples of the dataset.
4. The ALOMLP try to construct the fittest MLP with minimum MSE.
5. Update the ants: the ants are updated.
6. Repeat steps 2 to 4 till the last iteration.
7. Conclusion and testing: lastly, the training is completed and we test the obtained MLP based on the validation instances.

The overall steps of the ALOMLP are shown in Fig. 3.

6 Results and Discussions

In this section, the proposed ALOMLP approach is applied to 16 datasets with various number features and instances, which are obtained from UCI Machine Learning Repository [62]. Table 1 shows the details of the used datasets. All experiments were performed using Matlab 7.10 (R2010a) on a 64-bit Windows server 2012 with 1.70 Ghz CPU and 64 GB RAM. No commercial ALO-based utility was employed in this research. Detail of optimizer's parameters shown in Table 2.

Table 3 shows the average accuracy results and standard deviation (STD) of ALO in comparison with those of other optimizers. Table 4 tabulates the ranking results of all optimizers based on the results in Table 3. From Table 3, it can be detected

Table 1 Summary of the classification data sets

Data set	#Features	#Training samples	#Testing samples
Australian	14	455	235
Bank	48	2261	2260
Breast	8	461	238
Chess	36	2109	1087
Credit	61	670	330
Ionosphere	33	231	120
Mammographic-mass	5	633	328
Monk	6	285	147
Phoneme	5	3566	1838
Ring	20	4884	2516
Sonar	60	137	71
Spam base	57	2300	2301
Tic-Tac-Toe	9	632	326
Titanic	3	1452	749
Twonorm	20	4884	2516
WDBC	30	375	194

Table 2 The initial settings of the parameters in the metaheuristic algorithms

Algorithm	Parameter	Value
GA	Crossover probability	0.9
	Mutation probability	0.1
	Selection mechanism	Stochastic Universal Sampling
	Population size	50
	Maximum generations	200
PSO	Acceleration constants	[2.1,2.1]
	Intertia weights	[0.9,0.6]
	Number of particles	50
DE	Crossover probability	0.9
	Differential weight	0.5
	Population size	50
PBIL	Learning rate	0.05
	Good population member	1
	Bad population member	0
	Elitism parameter	1
	Mutational probability	0.1

that ALO can outperform the well-known DE, GA, PSO, and PBIL on austrilian, chess, ionosphere, mammographic-mass, phoneme, ring, sonar, and tictac datasets. While ALO is outperforming other peers on 50% of cases, it can be observed that GA has attained the second best rank and provides satisfactory accuracy results on 37.5 % of datasets. As per ranking marks in Table 4, after ALO and GA, wee see that PBIL, PSO, and DE algorithms are the next alternatives, respectively. The well-established PBIL can provide slightly better results than other peers on credit and titanic datasets, whereas the results of all methods are very competitive. On the other hand, PSO and DE could not attain any best solution for all datasets. The evolutionary convergence curves of ALO, DE, PSO, GA, and PBIL in tackling all datasets are compared in Fig. 5. If we focus on then convergence behaviors in Fig. 5, when inspecting the curves in Fig. 5c, we detect that even with better solutions of GA on breast dataset in Table 3, the novel ALO still surpass well-tuned GA in convergence rate. Same observation is verified on monk, spambase, and wdbc datasets. The core reason is that the ALO is equipped with time-varying dwindling edges of antlions cones and roulette wheel selection, while GA only utilize the former operation. Based on curves in Fig. 5, we realize that ALO demonstrates the fastest behavior is spite of DE algorithm. In addition, stagnation drawback of PSO is also detectable in Fig. 5d, j, m.

Table 3 Comparison of average accuracy results of ALO versus other competitors

Algorithms	ALO		DE		GA		PSO		PBIL	
Datasets	AVG.	STD	AVG.	STD	AVG.	STD	AVG.	STD	AVG.	STD
Austrilian	**0.8405**	0.017713	0.75877	0.051019	0.82047	0.023034	0.81784	0.021816	0.82544	0.015384
Bank	0.88786	0.00245	0.88503	0.004733	**0.8937**	0.002724	0.88543	0.00473	0.88792	0.003688
Breast	0.97213	0.006191	0.93908	0.019638	**0.97227**	0.006203	0.96457	0.009788	0.96611	0.007797
Chess	**0.74627**	0.030705	0.62628	0.028318	0.6149	0.077092	0.68847	0.025671	0.70506	0.022704
Credit	0.70061	0.019385	0.69172	0.021278	0.70111	0.016853	0.69889	0.021361	**0.70687**	0.021278
Ionosphere	**0.78611**	0.034625	0.73306	0.056157	0.75556	0.042398	0.76139	0.043904	0.77972	0.034302
Mammographic-mass	**0.79583**	0.007359	0.7939	0.017108	0.79339	0.005412	0.79167	0.011694	0.78821	0.013247
Monk	0.80295	0.018348	0.71202	0.055621	**0.80794**	0.021114	0.77324	0.034333	0.77324	0.0341
Phoneme	**0.76017**	0.006783	0.73946	0.024508	0.75851	0.011346	0.75637	0.009628	0.75843	0.015969
Ring	**0.75073**	0.012334	0.64988	0.022269	0.69899	0.030148	0.69047	0.025963	0.71241	0.01588
Sonar	**0.68638**	0.04101	0.61502	0.072134	0.58685	0.07538	0.62911	0.067225	0.6723	0.059403
Spambase	0.76015	0.033074	0.65661	0.04878	**0.76555**	0.02499	0.73325	0.027484	0.7322	0.028306
Tictac	**0.66534**	0.019869	0.60695	0.026714	0.63231	0.01994	0.63037	0.033716	0.64417	0.027436
Titanic	0.76204	0.003339	0.76226	0.007649	0.76213	0.005033	0.76351	0.00626	**0.76462**	0.006691
Twonorm	0.95146	0.008643	0.80624	0.04021	**0.97352**	0.002271	0.90391	0.019633	0.92565	0.013194
WDBC	0.93247	0.021203	0.86168	0.037857	**0.9445**	0.017789	0.92698	0.018515	0.92457	0.013434

Table 4 Ranking of average accuracy results for all algorithms on all datasets

Datasets	ALO	DE	GA	PSO	PBIL
Austrilian	1	5	3	4	2
Bank	3	5	1	4	2
Breast	2	5	1	4	3
Chess	1	4	5	3	2
Credit	3	5	2	4	1
Ionosphere	1	5	4	3	2
Mammographic-mass	1	2	3	4	5
Monk	2	5	1	3	3
Phoneme	1	5	2	4	3
Ring	1	5	3	4	2
Sonar	1	4	5	3	2
Spambase	2	5	1	3	4
Tictac	1	5	3	4	2
Titanic	5	3	4	2	1
Twonorm	2	5	1	4	3
WDBC	2	5	1	3	4
Avergae Rank	1.8125	4.5625	2.5000	3.5000	2.5625

Table 5 reveals the best, worst and median of the accuracy results for all competitors on all datasets. Figure 4 also shows the boxplots of these results for specific datasets tackled by ALO, efficiently. According to the best, worst and median of the accuracy results in Table 2, we recognize that ALO is still superior to other peers in solving the majority of datasets. For instance, new ALO has improved the median of the MSE results up to 0.9% compared to the second best optimizer, which is PBIL. The exposed boxplots in Fig. 4 also confirm the better-quality solutions (MLPs) that is realized by ALO on austrilian, chess, ionosphere, mammographic-mass, phoneme, ring, sonar, and tictac datasets.

To recapitulate, the overall results in terms of best, worst, median, and average of accuracy values divulge that the ALO has effectively enriched the classification accuracy of the MLP network. The performance of ALO-based MLP network does not deteriorate based on total number of features, training, and test samples. This fact is owing to the flexibility of ALO in handling a large number of candidate MLPs. The ALO has an elite-based evolutionary scheme, tournament selection, random walk, and adaptive parameters. These merits allow ALO to evade immature convergence shortcomings and inertia to LO.

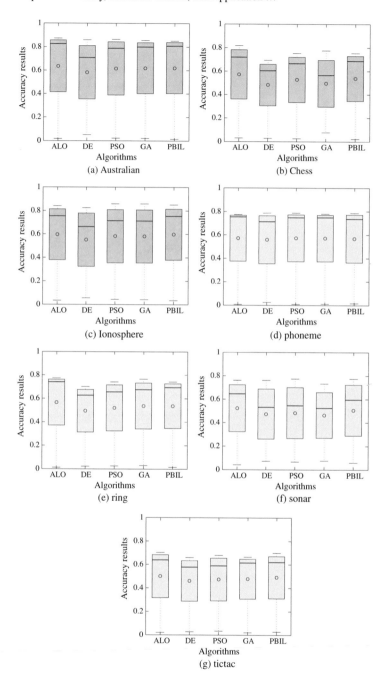

Fig. 4 Boxplots for seven datasets that ALO outperforms other competitors

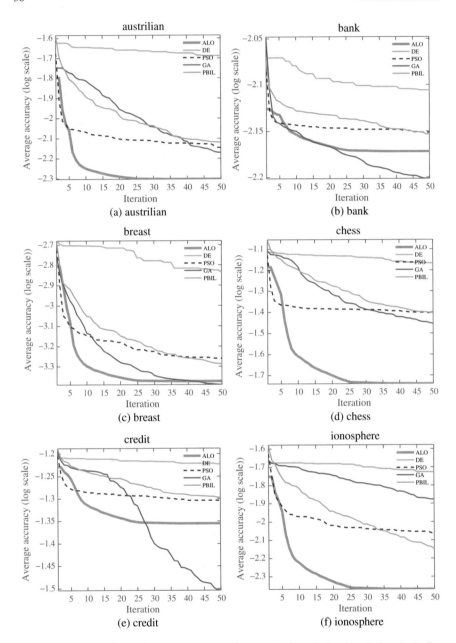

Fig. 5 Visual comparison of convergence curves in logarithmic scale for all optimizer in dealing with all 16 datasets

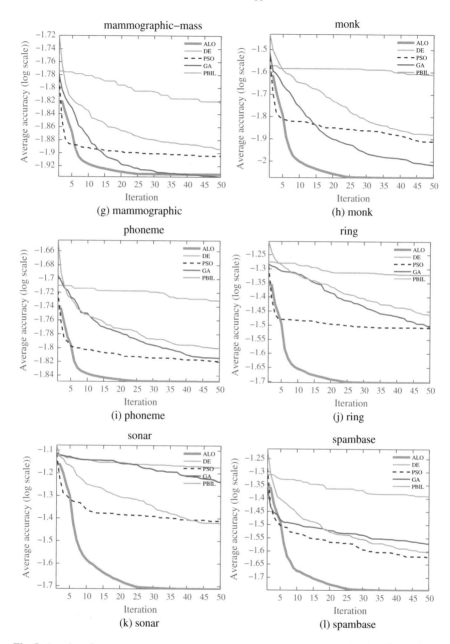

(g) mammographic

(h) monk

(i) phoneme

(j) ring

(k) sonar

(l) spambase

Fig. 5 (continued)

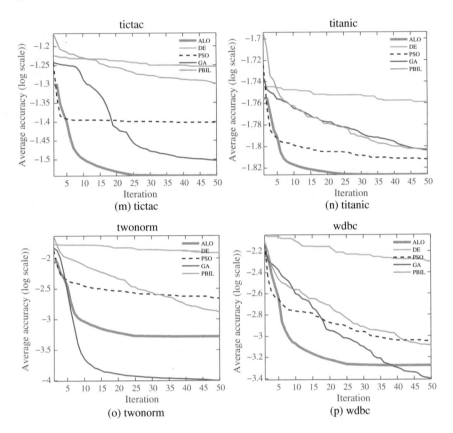

Fig. 5 (continued)

7 Conclusions and Future Directions

This book chapter was devoted to the theoretical backgrounds of well-established ALO, reviewing the previous works on ALO, comprehensively, and then, proposing a new integrated ALOMLP approach. After presenting an encoding scheme and defining the objective formula, the ALOMLP was utilized to deal with sixteen datasets. After presenting an encoding scheme and defining the objective formula, the ALOMLP was utilized to deal with sixteen datasets. The performance of ALO was compared to PSO, DE, GA, and PBIL in terms of best, worst, average, and median accuracy. In addition, the convergence behaviors were monitored for all methods. The comparative results expose that the ALOMLP classifies majority of datasets with best performance, efficiently.

Table 5 Minimum (MIN), maximum (MAX) and median (MED.) of accuracy results for all algorithms for used datasets

Algorithms	ALO			DE			GA			PSO			PBIL		
Dataset	MAX	MIN	MED.	MAX	MIN	MED.	MAX	MIN	MED.	MAX	MIN	MED.	MAX	MIN	MED.
Austrilian	0.87281	0.8114	0.83772	0.85965	0.65789	0.76316	0.86404	0.75439	0.82237	0.85526	0.7807	0.81579	0.85088	0.78947	0.82895
Bank	0.89381	0.88363	0.88783	0.89027	0.86681	0.88628	0.89912	0.88584	0.89403	0.89381	0.87301	0.88695	0.89381	0.87611	0.88783
Breast	0.98319	0.95798	0.97059	0.97059	0.89496	0.94118	0.98319	0.95798	0.97479	0.97899	0.94538	0.96639	0.97899	0.95378	0.96639
Chess	0.81877	0.69273	0.74241	0.69365	0.58142	0.62098	0.77461	0.51334	0.59982	0.75253	0.64121	0.68951	0.75253	0.66697	0.69917
Credit	0.73636	0.65455	0.7	0.73939	0.64545	0.69242	0.71818	0.64242	0.70909	0.73636	0.63636	0.7	0.75758	0.66667	0.70606
Ionosphere	0.84167	0.725	0.78333	0.825	0.59167	0.74167	0.85833	0.66667	0.75833	0.85833	0.66667	0.75833	0.85	0.725	0.77917
Mammographic-mass	0.81707	0.78354	0.79573	0.81707	0.75	0.79878	0.80793	0.78354	0.79268	0.81707	0.76829	0.78963	0.80488	0.75	0.79268
Monk	0.85034	0.76871	0.80272	0.83673	0.59184	0.72109	0.84354	0.76871	0.80272	0.85034	0.70068	0.76871	0.85034	0.70748	0.78231
Phoneme	0.77312	0.74483	0.75952	0.78672	0.68498	0.74129	0.77856	0.73613	0.75925	0.78618	0.73667	0.75707	0.78455	0.71328	0.75952
Ring	0.77424	0.73052	0.7498	0.70191	0.60413	0.65163	0.7651	0.65223	0.69237	0.74126	0.62202	0.69436	0.74205	0.67647	0.71522
Sonar	0.76056	0.60563	0.69014	0.76056	0.4507	0.61972	0.73239	0.46479	0.57746	0.77465	0.46479	0.64789	0.77465	0.52113	0.67606
Spambase	0.81443	0.69274	0.76793	0.73533	0.56063	0.66341	0.79226	0.69709	0.77466	0.78488	0.68144	0.73142	0.80139	0.67405	0.74055
Tictac	0.70245	0.6135	0.67025	0.65951	0.54908	0.60429	0.66564	0.59816	0.6365	0.68098	0.54908	0.63344	0.69939	0.59509	0.63957
Titanic	0.76903	0.75968	0.76101	0.77036	0.73832	0.76101	0.77036	0.75167	0.76101	0.77036	0.74633	0.76769	0.77837	0.73965	0.76569
Twonorm	0.96542	0.93243	0.95231	0.87599	0.69356	0.80346	0.97774	0.9686	0.97337	0.94515	0.86208	0.90819	0.95191	0.90143	0.92548
WDBC	0.97423	0.87629	0.93299	0.93299	0.76804	0.85825	0.98454	0.89691	0.9433	0.96907	0.88144	0.92784	0.95361	0.90206	0.92526

References

1. Abbassi, R., Abbassi, A., Heidari, A. A., & Mirjalili, S. (2019). An efficient salp swarm-inspired algorithm for parameters identification of photovoltaic cell models. *Energy Conversion and Management, 179*, 362–372.
2. Ahmad, S., Mafarja, M., Faris, H., & Aljarah, I. (2018). Feature selection using salp swarm algorithm with chaos.
3. Alba, E., & Chicano, J. (2004). Training neural networks with ga hybrid algorithms. In *Genetic and Evolutionary Computation–GECCO 2004* (pp. 852–863). Springer.
4. Ali, E., Elazim, S. A., & Abdelaziz, A. (2017). Ant lion optimization algorithm for optimal location and sizing of renewable distributed generations. *Renewable Energy, 101*, 1311–1324.
5. Ali, E., Elazim, S. A., & Abdelaziz, A. (2018). Optimal allocation and sizing of renewable distributed generation using ant lion optimization algorithm. *Electrical Engineering, 100*(1), 99–109.
6. Aljarah, I., AlaM, A. Z., Faris, H., Hassonah, M. A., Mirjalili, S., & Saadeh, H. (2018). Simultaneous feature selection and support vector machine optimization using the grasshopper optimization algorithm. *Cognitive Computation* (pp. 1–18).
7. Aljarah, I., Faris, H., & Mirjalili, S. (2016). Optimizing connection weights in neural networks using the whale optimization algorithm. *Soft Computing* (pp. 1–15).
8. Aljarah, I., Faris, H., Mirjalili, S., & Al-Madi, N. (2018). Training radial basis function networks using biogeography-based optimizer. *Neural Computing and Applications, 29*(7), 529–553.
9. Aljarah, I., & Ludwig, S. A. (2012). Parallel particle swarm optimization clustering algorithm based on mapreduce methodology. In *Proceedings of the Fourth World Congress on Nature and Biologically Inspired Computing (IEEE NaBIC12)*. IEEE Explore.
10. Aljarah, I., & Ludwig, S. A. (2013). A new clustering approach based on glowworm swarm optimization. In *Proceedings of 2013 IEEE Congress on Evolutionary Computation Conference (IEEE CEC13)*, Cancun, Mexico. IEEE Xplore.
11. Aljarah, I., Mafarja, M., Heidari, A. A., Faris, H., Zhang, Y., & Mirjalili, S. (2018). Asynchronous accelerating multi-leader salp chains for feature selection. *Applied Soft Computing, 71*, 964–979.
12. Almonacid, F., Fernandez, E. F., Mellit, A., & Kalogirou, S. (2017). Review of techniques based on artificial neural networks for the electrical characterization of concentrator photovoltaic technology. *Renewable and Sustainable Energy Reviews, 75*, 938–953.
13. Ata, R. (2015). Artificial neural networks applications in wind energy systems: a review. *Renewable and Sustainable Energy Reviews, 49*, 534–562.
14. Aminisharifabad, M., Yang, Q. & Wu, X. (2018). A penalized Autologistic regression with application for modeling the microstructure of dual-phase high strength steel. *Journal of Quality Technology*. in-press.
15. Blum, C., & Socha, K. (2005). Training feed-forward neural networks with ant colony optimization: An application to pattern classification. In *Fifth International Conference on Hybrid Intelligent Systems, 2005. HIS'05* (p. 6). IEEE.
16. Braik, M., Sheta, A., & Arieqat, A.: (2008). A comparison between gas and pso in training ann to model the te chemical process reactor. In *AISB 2008 Convention Communication, Interaction And Social Intelligence* (vol. 1, p. 24).
17. Cao, W., Yan, C., Wu, D., & Tuo, J. (2017). A novel multi-objective optimization approach of machining parameters with small sample problem in gear hobbing. *The International Journal of Advanced Manufacturing Technology, 93*(9–12), 4099–4110.
18. Chaudhuri, B., & Bhattacharya, U. (2000). Efficient training and improved performance of multilayer perceptron in pattern classification. *Neurocomputing, 34*(1), 11–27.
19. Chen, J. F., Do, Q. H., & Hsieh, H. N. (2015). Training artificial neural networks by a hybrid pso-cs algorithm. *Algorithms, 8*(2), 292–308.
20. Chitsaz, H., & Aminisharifabad, M. (2015). Exact learning of rna energy parameters from structure. *Journal of Computational Biology, 22*(6), 463–473.

21. Cybenko, G. (1989). Approximation by superpositions of a sigmoidal function. *Mathematics of Control, Signals, and Systems (MCSS), 2*(4), 303–314.
22. Ding, S., Li, H., Su, C., Yu, J., & Jin, F. (2013). Evolutionary artificial neural networks: A review. *Artificial Intelligence Review* (pp. 1–10).
23. Dinkar, S. K., & Deep, K. (2017). Opposition based laplacian ant lion optimizer. *Journal of Computational Science, 23*, 71–90.
24. Dubey, H. M., Pandit, M., & Panigrahi, B. (2016). Ant lion optimization for short-term wind integrated hydrothermal power generation scheduling. *International Journal of Electrical Power & Energy Systems, 83*, 158–174.
25. Dubey, H. M., Pandit, M., & Panigrahi, B. (2018). An overview and comparative analysis of recent bio-inspired optimization techniques for wind integrated multi-objective power dispatch. *Swarm and Evolutionary Computation, 38*, 12–34.
26. Elaziz, M. A., Moemen, Y. S., Hassanien, A. E., & Xiong, S. (2018). Quantitative structure-activity relationship model for hcvns5b inhibitors based on an antlion optimizer-adaptive neuro-fuzzy inference system. *Scientific reports, 8*(1), 1506.
27. Emary, E., Zawbaa, H. M., & Hassanien, A. E. (2016). Binary ant lion approaches for feature selection. *Neurocomputing, 213*, 54–65.
28. Esteva, A., Kuprel, B., Novoa, R. A., Ko, J., Swetter, S. M., Blau, H. M., et al. (2017). Dermatologist-level classification of skin cancer with deep neural networks. *Nature, 542*(7639), 115–118.
29. Faris, H., Ala'M, A. Z., Heidari, A. A., Aljarah, I., Mafarja, M., Hassonah, M. A., & Fujita, H. (2019). An intelligent system for spam detection and identification of the most relevant features based on evolutionary random weight networks. *Information Fusion, 48*, 67–83.
30. Faris, H., Aljarah, I., Al-Betar, M.A., & Mirjalili, S. (2017). Grey wolf optimizer: A review of recent variants and applications. *Neural Computing and Applications, 1–23*.
31. Faris, H., Aljarah, I., Al-Madi, N., & Mirjalili, S. (2016). Optimizing the learning process of feedforward neural networks using lightning search algorithm. *International Journal on Artificial Intelligence Tools, 25*(06), 1650033.
32. Faris, H., Aljarah, I., & Al-Shboul, B. (2016). A hybrid approach based on particle swarm optimization and random forests for e-mail spam filtering. In *International Conference on Computational Collective Intelligence* (pp. 498–508). Springer, Cham.
33. Faris, H., Aljarah, I., & Mirjalili, S. (2016). Training feedforward neural networks using multiverse optimizer for binary classification problems. *Applied Intelligence, 45*(2), 322–332.
34. Faris, H., Aljarah, I., & Mirjalili, S. (2017). Evolving radial basis function networks using moth–flame optimizer. In *Handbook of Neural Computation* (pp. 537–550).
35. Faris, H., Aljarah, I., & Mirjalili, S. (2017). Improved monarch butterfly optimization for unconstrained global search and neural network training. *Applied Intelligence* (pp. 1–20).
36. Faris, H., & Aljarah, I., et al. (2015). Optimizing feedforward neural networks using krill herd algorithm for e-mail spam detection. In *2015 IEEE Jordan Conference on Applied Electrical Engineering and Computing Technologies (AEECT)* (pp. 1–5). IEEE.
37. Faris, H., Hassonah, M. A., AlaM, A. Z., Mirjalili, S., & Aljarah, I. (2017). A multi-verse optimizer approach for feature selection and optimizing svm parameters based on a robust system architecture. *Neural Computing and Applications, 1–15*.
38. Faris, H., Mafarja, M. M., Heidari, A. A., Aljarah, I., AlaM, A. Z., Mirjalili, S., et al. (2018). An efficient binary salp swarm algorithm with crossover scheme for feature selection problems. *Knowledge-Based Systems, 154*, 43–67.
39. Gandomi, A. H., & Kashani, A. R. (2018). Construction cost minimization of shallow foundation using recent swarm intelligence techniques. *IEEE Transactions on Industrial Informatics, 14*(3), 1099–1106.
40. Green, R. C., Wang, L., & Alam, M. (2012). Training neural networks using central force optimization and particle swarm optimization: insights and comparisons. *Expert Systems with Applications, 39*(1), 555–563.
41. Gupta, S., Kumar, V., Rana, K., Mishra, P., & Kumar, J. (2016). Development of ant lion optimizer toolkit in labview. In *2016 International Conference onInnovation and Challenges in Cyber Security (ICICCS-INBUSH)* (pp. 251–256).

42. Hamouda, E., El-Metwally, S., & Tarek, M. (2018). Ant lion optimization algorithm for kidney exchanges. *PloS One, 13*(5), e0196707.
43. Hansen, N., Müller, S. D., & Koumoutsakos, P. (2003). Reducing the time complexity of the derandomized evolution strategy with covariance matrix adaptation (cma-es). *Evolutionary Computation, 11*(1), 1–18.
44. Heidari, A. A., Faris, H., Aljarah, I., & Mirjalili, S. (2018). An efficient hybrid multilayer perceptron neural network with grasshopper optimization. *Soft Computing*, 1–18.
45. Heidari, A. A., Abbaspour, R. A. (2018). Enhanced chaotic grey wolf optimizer for real-world optimization problems: A comparative study. In *Handbook of Research on Emergent Applications of Optimization Algorithms* (pp. 693–727). IGI Global.
46. Heidari, A. A., Abbaspour, R. A., & Jordehi, A. R. (2017). An efficient chaotic water cycle algorithm for optimization tasks. *Neural Computing and Applications, 28*(1), 57–85.
47. Heidari, A. A., Abbaspour, R. A., & Jordehi, A. R. (2017). Gaussian bare-bones water cycle algorithm for optimal reactive power dispatch in electrical power systems. *Applied Soft Computing, 57*, 657–671.
48. Heidari, A. A., & Delavar, M. R. (2016). A modified genetic algorithm for finding fuzzy shortest paths in uncertain networks. *ISPRS - International Archives of the Photogrammetry, Remote Sensing and Spatial Information Sciences XLI-B2* (pp. 299–304).
49. Heidari, A. A., & Pahlavani, P. (2017). An efficient modified grey wolf optimizer with lévy flight for optimization tasks. *Applied Soft Computing, 60*, 115–134.
50. Ho, Y. C., & Pepyne, D. L. (2002). Simple explanation of the no free lunch theorem of optimization. *Cybernetics and Systems Analysis, 38*(2), 292–298.
51. Hu, P., Wang, Y., Wang, H., Zhao, R., Yuan, C., Zheng, Y., Lu, Q., Li, Y., & Masood, I. (2018). Alo-dm: A smart approach based on ant lion optimizer with differential mutation operator in big data analytics. In *International Conference on Database Systems for Advanced Applications* (pp. 64–73). Springer.
52. Hu, Y. C. (2014). Nonadditive similarity-based single-layer perceptron for multi-criteria collaborative filtering. *Neurocomputing, 129*, 306–314.
53. Igel, C., & Toussaint, M. (2003). On classes of functions for which no free lunch results hold. *Information Processing Letters, 86*(6), 317–321.
54. Ilonen, J., Kamarainen, J. K., & Lampinen, J. (2003). Differential evolution training algorithm for feed-forward neural networks. *Neural Processing Letters, 17*(1), 93–105.
55. Kamboj, V. K., Bhadoria, A., & Bath, S. (2017). Solution of non-convex economic load dispatch problem for small-scale power systems using ant lion optimizer. *Neural Computing and Applications, 28*(8), 2181–2192.
56. Karaboga, D., Akay, B., & Ozturk, C. (2007). Artificial bee colony (abc) optimization algorithm for training feed-forward neural networks. In *International Conference on Modeling Decisions for Artificial Intelligence* (pp. 318–329). Springer.
57. Kaushal, K., & Singh, S. (2017). Allocation of stocks in a portfolio using antlion algorithm: Investor's perspective. *IUP Journal of Applied Economics, 16*(1), 34.
58. Kowalski, P. A., & Łukasik, S. (2016). Training neural networks with krill herd algorithm. *Neural Processing Letters, 44*(1), 5–17.
59. Krogh, A. (2008). What are artificial neural networks? *Nature Biotechnology, 26*(2), 195–197.
60. Lee, S., & Choeh, J. Y. (2014). Predicting the helpfulness of online reviews using multilayer perceptron neural networks. *Expert Systems with Applications, 41*(6), 3041–3046.
61. Li, Y., Feng, B., Li, G., Qi, J., Zhao, D., & Mu, Y. (2018). Optimal distributed generation planning in active distribution networks considering integration of energy storage. *Applied Energy, 210*, 1073–1081.
62. Lichman, M.: UCI machine learning repository (2013), http://archive.ics.uci.edu/ml
63. Mafarja, M., Aljarah, I., Heidari, A. A., Faris, H., Fournier-Viger, P., Li, X., & Mirjalili, S. (2018). Binary dragonfly optimization for feature selection using time-varying transfer functions. *Knowledge-Based Systems, 161*, 185–204.
64. Mafarja, M., Aljarah, I., Heidari, A. A., Hammouri, A. I., Faris, H., & AlaM, A. Z., et al. (2017). Evolutionary population dynamics and grasshopper optimization approaches for feature selection problems. *Knowledge-Based Systems*.

65. Mallipeddi, R., Suganthan, P. N., Pan, Q. K., & Tasgetiren, M. F. (2011). Differential evolution algorithm with ensemble of parameters and mutation strategies. *Applied Soft Computing, 11*(2), 1679–1696.
66. McCulloch, W. S., & Pitts, W. (1943). A logical calculus of the ideas immanent in nervous activity. *The Bulletin of Mathematical Biophysics, 5*(4), 115–133.
67. Mirjalili, S. (2015). The ant lion optimizer. *Advances in Engineering Software, 83*, 80–98.
68. Mirjalili, S. (2015). How effective is the grey wolf optimizer in training multi-layer perceptrons. *Applied Intelligence, 43*(1), 150–161.
69. Mirjalili, S., Jangir, P., & Saremi, S. (2017). Multi-objective ant lion optimizer: A multi-objective optimization algorithm for solving engineering problems. *Applied Intelligence, 46*(1), 79–95.
70. Mirjalili, S. Z., Mirjalili, S., Saremi, S., Faris, H., & Aljarah, I. (2018). Grasshopper optimization algorithm for multi-objective optimization problems. *Applied Intelligence, 48*(4), 805–820.
71. Mirjalili, S. Z., Saremi, S., & Mirjalili, S. M. (2015). Designing evolutionary feedforward neural networks using social spider optimization algorithm. *Neural Computing and Applications, 26*(8), 1919–1928.
72. Nair, S. S., Rana, K., Kumar, V., & Chawla, A. (2017). Efficient modeling of linear discrete filters using ant lion optimizer. *Circuits, Systems, and Signal Processing, 36*(4), 1535–1568.
73. Ojha, V. K., Abraham, A., & Snášel, V. (2017). Metaheuristic design of feedforward neural networks: A review of two decades of research. *Engineering Applications of Artificial Intelligence, 60*, 97–116.
74. Oliva, D., Hinojosa, S., Elaziz, M.A., & Ortega-Sánchez, N. (2018). Context based image segmentation using antlion optimization and sine cosine algorithm. *Multimedia Tools and Applications* (pp. 1–37).
75. Petrović, M., Petronijević, J., Mitić, M., Vuković, N., Miljković, Z., & Babić, B. (2016). The ant lion optimization algorithm for integrated process planning and scheduling. In *Applied Mechanics and Materials* (vol. 834, pp. 187–192). Trans Tech Publ.
76. Rajan, A., Jeevan, K., & Malakar, T. (2017). Weighted elitism based ant lion optimizer to solve optimum var planning problem. *Applied Soft Computing, 55*, 352–370.
77. Raju, M., Saikia, L. C., & Sinha, N. (2016). Automatic generation control of a multi-area system using ant lion optimizer algorithm based pid plus second order derivative controller. *International Journal of Electrical Power & Energy Systems, 80*, 52–63.
78. Saxena, P., & Kothari, A. (2016). Ant lion optimization algorithm to control side lobe level and null depths in linear antenna arrays. *AEU-International Journal of Electronics and Communications, 70*(9), 1339–1349.
79. Schumacher, C., Vose, M. D., & Whitley, L. D. (2001). The no free lunch and problem description length. In *Proceedings of the 3rd Annual Conference on Genetic and Evolutionary Computation* (pp. 565–570). Morgan Kaufmann Publishers Inc.
80. Seeley, W. W., Crawford, R. K., Zhou, J., Miller, B. L., & Greicius, M. D. (2009). Neurodegenerative diseases target large-scale human brain networks. *Neuron, 62*(1), 42–52.
81. Sexton, R. S., Dorsey, R. E., & Johnson, J. D. (1999). Optimization of neural networks: A comparative analysis of the genetic algorithm and simulated annealing. *European Journal of Operational Research, 114*(3), 589–601.
82. Sexton, R. S., & Gupta, J. N. (2000). Comparative evaluation of genetic algorithm and backpropagation for training neural networks. *Information Sciences, 129*(1), 45–59.
83. Shukri, S., Faris, H., Aljarah, I., Mirjalili, S., & Abraham, A. (2018). Evolutionary static and dynamic clustering algorithms based on multi-verse optimizer. *Engineering Applications of Artificial Intelligence, 72*, 54–66.
84. Siddique, M., & Tokhi, M. (2001) Training neural networks: backpropagation vs. genetic algorithms. In *International Joint Conference on Neural Networks, 2001. Proceedings. IJCNN'01* (vol. 4, pp. 2673–2678). IEEE.
85. Slowik, A., & Bialko, M. (2008). Training of artificial neural networks using differential evolution algorithm. In *2008 Conference on Human System Interactions* (pp. 60–65). IEEE.

86. Socha, K., & Blum, C. (2007). An ant colony optimization algorithm for continuous optimization: application to feed-forward neural network training. *Neural Computing and Applications*, *16*(3), 235–247.
87. Talatahari, S. (2016). Optimum design of skeletal structures using ant lion optimizer. *Iran University of Science & Technology*, *6*(1), 13–25.
88. Tharwat, A., & Hassanien, A. E. (2018). Chaotic antlion algorithm for parameter optimization of support vector machine. *Applied Intelligence*, *48*(3), 670–686.
89. Tian, T., Liu, C., Guo, Q., Yuan, Y., Li, W., & Yan, Q. (2018). An improved ant lion optimization algorithm and its application in hydraulic turbine governing system parameter identification. *Energies, 11*(1), 95.
90. Trivedi, I. N., Jangir, P., & Parmar, S. A. (2016). Optimal power flow with enhancement of voltage stability and reduction of power loss using ant-lion optimizer. *Cogent Engineering, 3*(1), 1208942.
91. Trujillo, M. C. R., Alarcón, T. E., Dalmau, O. S., & Ojeda, A. Z. (2017). Segmentation of carbon nanotube images through an artificial neural network. *Soft Computing, 21*(3), 611–625.
92. Wdaa, A. S. I. (2008). Differential evolution for neural networks learning enhancement. Ph.D. thesis, Universiti Teknologi Malaysia.
93. Whitley, D., Starkweather, T., & Bogart, C. (1990). Genetic algorithms and neural networks: Optimizing connections and connectivity. *Parallel Computing, 14*(3), 347–361.
94. Wienholt, W. (1993). Minimizing the system error in feedforward neural networks with evolution strategy. In *ICANN93* (pp. 490–493). Springer.
95. Wolpert, D. H., & Macready, W. G. (1997). No free lunch theorems for optimization. *IEEE Transactions on Evolutionary Computation, 1*(1), 67–82.
96. Wu, Z., Yu, D., & Kang, X. (2017). Parameter identification of photovoltaic cell model based on improved ant lion optimizer. *Energy Conversion and Management, 151*, 107–115.
97. Yamany, W., Tharwat, A., Hassanin, M. F., Gaber, T., Hassanien, A.E., & Kim, T. H. (2015). A new multi-layer perceptrons trainer based on ant lion optimization algorithm. In *2015 Fourth International Conference on Information Science and Industrial Applications (ISI)* (pp. 40–45). IEEE.
98. Yao, P., & Wang, H. (2017). Dynamic adaptive ant lion optimizer applied to route planning for unmanned aerial vehicle. *Soft Computing, 21*(18), 5475–5488.
99. Yogarajan, G., & Revathi, T. (2018). Improved cluster based data gathering using ant lion optimization in wireless sensor networks. *Wireless Personal Communications, 98*(3), 2711–2731.
100. Yu, J., Wang, S., & Xi, L. (2008). Evolving artificial neural networks using an improved pso and dpso. *Neurocomputing, 71*(4), 1054–1060.
101. Zawbaa, H. M., Emary, E., & Grosan, C. (2016). Feature selection via chaotic antlion optimization. *PloS One, 11*(3), e0150652.
102. Zhang, J. R., Zhang, J., Lok, T. M., & Lyu, M. R. (2007). A hybrid particle swarm optimization-back-propagation algorithm for feedforward neural network training. *Applied Mathematics and Computation, 185*(2), 1026–1037.

Dragonfly Algorithm: Theory, Literature Review, and Application in Feature Selection

Majdi Mafarja, Ali Asghar Heidari, Hossam Faris, Seyedali Mirjalili and Ibrahim Aljarah

Abstract In this chapter, a wrapper-based feature selection algorithm is designed and substantiated based on the binary variant of Dragonfly Algorithm (BDA). DA is a successful, well-established metaheuristic that revealed superior efficacy in dealing with various optimization problems including feature selection. In this chapter we are going first present the inspirations and methamatical modeds of DA in details. Then, the performance of this algorithm is tested on a special type of datasets that contain a huge number of features with low number of samples. This type of datasets makes the optimization process harder, because of the large search space, and the lack of adequate samples to train the model. The experimental results showed the ability of DA to deal with this type of datasets better than other optimizers in the literature. Moreover, an extensive literature review for the DA is provided in this chapter.

M. Mafarja
Faculty of Engineering and Technology, Department of Computer Science,
Birzeit University, PoBox 14, Birzeit, Palestine
e-mail: mmafarja@birzeit.edu

A. A. Heidari
School of Surveying and Geospatial Engineering, University of Tehran, Tehran, Iran
e-mail: as_heidari@ut.ac.ir

H. Faris · I. Aljarah
King Abdullah II School for Information Technology, The University of Jordan,
Amman, Jordan
e-mail: hossam.faris@ju.edu.jo

I. Aljarah
e-mail: i.aljarah@ju.edu.jo

S. Mirjalili (✉)
Institute of Integrated and Intelligent Systems, Griffith University, Nathan,
Brisbane, QLD 4111, Australia
e-mail: seyedali.mirjalili@griffithuni.edu.au

© Springer Nature Switzerland AG 2020
S. Mirjalili et al. (eds.), *Nature-Inspired Optimizers*, Studies in Computational
Intelligence 811, https://doi.org/10.1007/978-3-030-12127-3_4

47

1 Introduction

Some creatures' behaviors have been the inspiration source for many successful optimization algorithms. The main behavior that inspired many researchers to develop new algorithms was the strategy that those creatures use to seek the food sources. Ant Colony Optimization (ACO) [25, 26] and Artificial Bees Colony (ABC) [51] were originally inspired by the behavior of ants and bees respectively in locating food sources and collecting their food. Swarming behavior is another source of inspiration that was used to propose new optimization algorithms. Particle Swarm Optimization (PSO) [25] is a primary swarm based optimization algorithm that mimics the swarming behavior of birds.

The key issue about all the previously mentioned creatures is that they live in groups or folks (called swarms) [53]. An individual in those swarms usually makes a decision based on local information from itself and from the interactions with the other swarm members, also from the environment. Such interactions are the main reason that contribute to the improvement of the social intelligence in these swarms. Most of the swarms contain different organisms from the same species (bees, ants, birds, etc.). By the intelligent collaboration (or swarm intelligence (SI)) between all individuals of the swarm, they have been found to be successful in carrying out specific tasks [35, 95].

Nature-inspired algorithms are population-based metaheuristics algorithms, that tempt to manipulate a set of solutions in each generation of the optimization process [1, 13, 33, 44, 45, 53]. Recently, many such algorithms were proposed and proved their ability to solve different complex optimization problems such as global function optimization [12, 42–44, 46], clustering analysis [4, 8, 10, 90], spam and intrusion detection systems [9, 11, 30, 32], optimizing neural networks [6, 7, 29, 31], link prediction [18], software effort estimation [36], and bio-informatics [16, 23].

As Feature Selection is known to be an NP-hard problem [5], metaheuristic algorithms in general and specially the population-based algorithms showed a superior performance in solving this problem.

PSO has been widely used to tackle FS problems. An improved PSO with local search strategy for FS was proposed in [80]. Another PSO approach with two crossover mechanisms was presented in [22]. Rajamohana and Umamaheswari [82] proposed a hybrid particle swarm optimization with shuffled frog leaping algorithm feature selection approach. An improved ACO was also proposed in [52]. A hybrid FS approach that combines differential evolution (DE) with ABC was proposed in [102].

Grey Wolf Optimizer (GWO) [79] is a recent SI algorithm that mimics the hierarchical organization of the grey wolves in nature. The GWO has been widely used in FS methods with much success [58]. Antlion Optimizer (ALO) was proposed by Mirjalili [74] in 2015, FS was one of the fields where ALO was applied successfully [70]. Whale Optimization Algorithm (WOA), is another SI algorithm that was recently proposed by Mirjalili and Lewis [78]. WOA was successfully applied in many FS methods [72, 73]. Grasshopper Optimization Algorithm (GOA) [87] is another example of the nature inspired algorithms that was recently proposed and applied to many optimization problems including FS [5, 49, 69]. More recently, a Salp Swarm Algorithm (SSA) was proposed by Mirjalili et al. [76] and was used as a FS selection mechanism in [2].

Dragonfly Algorithm (DA) was recently proposed by Mirjalili [75]. DA is inspired by the behavior of dragonflies, and proved its ability outperform other well-regarded optimizers in the literature [68]. The DA has shown an excellent performance for several continuous, discrete, single-objective and multi-objective optimization problems compared to several state-of-the-art metaheuristic and evolutionary algorithms such as PSO and DE. In this chapter, the binary version of DA is used to serve as a selection mechanism and applied to medical datasets of high dimensional with small number of samples.

2 Feature Selection Problem

With the emergence of the high dimensional data in almost all of the real life fields, the knowledge discovery (KDD) process becomes very complex [61]. The high dimensionality of datasets usually causes many problems for the data mining techniques (e.g., classification), like over-fitting, and hence, decreasing the model performance, and increasing the required computational time. Therefore, reducing data dimensionality becomes highly demanded [59]. FS is a preprocessing step that aims to enhance the performance of the data mining and machine learning techniques by eliminating the redundant irrelevant features.

In the FS process, two major steps should be carefully considered when designing a new FS method; selection (search) and evaluation [61]. In the selection step, the feature space has to be searched for a feature subset with a minimal number of features, and reveals the highest model performance (e.g., classification accuracy when considering a classification technique). In the selection step, three different mechanisms can be considered; complete, random, and heuristic search mechanisms. When dealing with high-dimensional datasets, the complete search strategies (i.e., brute force) are impractical since 2^N feature subsets have to be generated and evaluated for a dataset with N features. The random mechanisms are also impractical to be used with FS methods since the search process is controlled by random factors that may lead the search process to be as worst as the complete search, in the worst case. Heuristic search algorithms are the most suitable mechanisms to be used with FS methods since the search process is guided by a heuristic that navigates the process towards finding the near-optimal solutions.

The evaluation criterion is another aspect that should be carefully considered when designing a FS method. In this regard, FS methods can be divided into three main types; filters, wrappers, and embedded methods. In filter approaches [63–66, 68], feature subsets are evaluated depending on the relations between features themselves, and their correlation with the class attribute without considering any particular classifier [62]. Thus, a feature with high score is selected to build the classification model in a further step. In the wrapper based methods [2, 34, 73], each feature subset is evaluated based on the performance of a particular classifier. Embedded methods (e.g., regularization) [20, 57, 61] learn which features enhance the performance of the model (e.g., accuracy), while the model is being created.

In summary, filter methods are considered as the fastest FS methods, since no external tools are required to be employed; nevertheless the model performance is uncertain. Wrappers are slower than filters, while they usually obtain higher accuracy. Embedded methods come in between filters and wrappers in terms of the model performance and the computational time.

3 Dragonfly Algorithm

Dragonfly Algorithm (DA) is a recently well-established population-based optimizer proposed by Mirjalili in 2016 [75]. The DA was developed based on the hunting and migration strategies of dragonflies. The hunting technique is known as static swarm (feeding), in which all members of a swarm can fly in small clusters over a limited

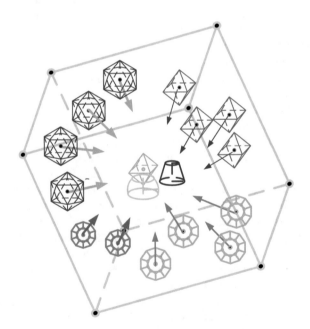

Fig. 1 Dynamic swarming behaviors (each geometric object shows a special type of agents)

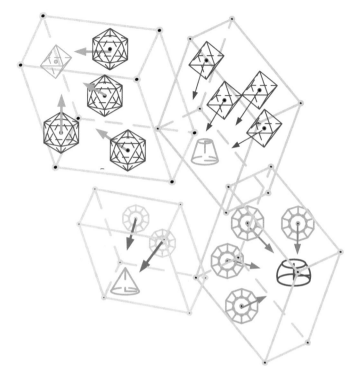

Fig. 2 Static swarming behaviors

space for discovering food sources. The migration strategy of dragonflies is called dynamic swarm (migratory). In this phase, the dragonflies are willing to soar in larger clusters, and as a result, the swarm can migrate. Dynamic and static groups are shown in Figs. 1 and 2. Likewise other swarm-based methods, the operators of DA perform two main concepts: diversification, motivated by the static swarming activities, and intensification, encouraged by the dynamic swarming activities.

In DA, five types of behaviors are designed as follows. In next parts, X is the position vector, X_j is the j-th neighbor of the X, and N denotes the neighborhood size [84]:

- *Separation* is a strategy that dragonflies use to separate themselves from other agents. This procedure is formulated as Eq. (1):

$$S_i = -\sum_{j=1}^{N} X - X_i \qquad (1)$$

- *Alignment* shows how an agent will set its velocity with respect to the velocity vector of other adjacent dragonflies. This concept is modeled based on Eq. (2):

$$A_i = \frac{\sum_{j=1}^{N} V_j}{N} \qquad (2)$$

where V_j indicates the velocity vector of the j-th neighbor.

- *Cohesion* indicates the inclination of members to move in the direction of the nearby center of mass. This step is formulated as in Eq. (3):

$$C_i = \frac{\sum_{j=1}^{N} x_j}{N} - X \qquad (3)$$

- *Attraction* shows to the propensity of members to move towards the food source. The attraction tendency among the food source and the ith agent is performed based on Eq. (4):

$$F_i = F_{loc} - X \qquad (4)$$

where F_{loc} is the food source's location.

- *Distraction* shows the proclivity of dragonflies to keep themselves away from an conflicting enemy. The distraction among the enemy and the ith dragonfly is performed according to Eq. (5):

$$E_i = E_{loc} + X \qquad (5)$$

where E_{loc} is the enemy's location.

In DA, the fitness of food source and position vectors are updated based on the fittest agent found so far. Moreover, the fitness values and positions of the enemy are computed based on the worst dragonfly. This fact can assist DA in converging towards more promising regions of the solution space and in turn, avoiding from non-promising areas. The position vectors of dragonflies are updated based on two rules: the step vector (ΔX) and the position vector. The step vector indicates the dragonflies' direction of motion and it is calculated as in Eq. (6):

$$\Delta X_{t+1} = (sS_i + aA_i + cC_i + fF_i + eE_i) + wX_t \qquad (6)$$

where s, w, a, c, f, and e show the weighting vectors of different components.

The location vector of members is calculated as in Eq. (7):

$$X_{t+1} = X_t + \Delta X_{t+1} \qquad (7)$$

where t is iteration.

Pseudocode of DA is shown in Algorithm 1.

Algorithm 1 Pseudocode of DA
***
Initialize the swarm $X_i (i = 1, 2, \ldots, n)$
Initialize $\Delta X_i (i = 1, 2, \ldots, n)$
while (end condition is not met) **do**
    Evaluate all dragonflies
    Update (F) and (E)
    Update coefficients $(i., e., w, s, a, c, f, and\ e)$
    Attain S, A, C, F, and E (based on Eqs. (9.21 to 4.5))
    Update step vectors (ΔX_{t+1}) by Eq. (4.6)
    Update X_{t+1} by Eq. (4.7)
Return the best agent
***

4 Literature Review

In this section, previous researches on DA and their core findings are reviewed in detail. From the time of proposing DA, several works have focused on the efficacy of DA or tried to improve its efficacy on tasks such as photovoltaic systems [83], extension of RFID network lifetime [47], economic emission dispatch (EPD) problems [19]. Hamdy et al. [39] used the multi-objective version of DA to solve the design problem of a nearly zero energy building (nZEB), along with six other evolutionary methods. In [85], a hybrid DA approach with extreme learning machine (ELM) was proposed. In this system, DA was used to optimize the number of nodes and their associated weights in the hidden layer of the ELM. Pathania et al. [81] proposed a DA based approach to solve the economic load dispatch (ELD) problem with valve point effect. DA was used in [24] to estimate the location of the unknown wireless nodes that are randomly deployed in a designated area. The binary version of DA was used with a multi-class SVM classifier within an FS method [28]. Hema Sekhar et al. [89] employed DA to optimize the firing angle and the size of the thyristor controlled series capacitor (TCSC). In [86], a self-adaptive DA was used to tackle the multilevel segmentation problem.

In 2017, a memory-based hybrid DA with particle swarm optimization (PSO) has been developed to deal with global optimization [55]. Song et al. [91] developed an enhanced DA with elite opposition learning (EOL) to tackle global optimization tasks. In [3], DA was used to solve the 0-1 knapsack problems. Another application that was solved using DA is the static economic dispatch incorporating solar energy as in [94]. DA was used as a FS search strategy in addition to tune the parameters of ELM in [101]. A multi objective version of DA (MODA) was used as an adaptive engine calibration algorithm to control the engine parameters [38]. Mafarja et al. [71] proposed a FS method that employed the binary version of DA (BDA) as a search strategy. In [96, 97], DA was used as a tool to optimize SVM parameters. In software engineering field, DA was also applied. In [93], DA was used as an optimization tool to select the most important test cases that satisfy all requirements. An improved version of DA based on elite opposition learning strategy was used to solve some global optimization problems [91].

Suresh and Sreejith [94] proposed the use of DA in optimizing the generation costs of all the participating units in a solar thermal economic dispatch system. Memory based Hybrid DA (MHDA) was proposed in [55] to solve numerical optimization problems. DA was used in [37] to optimize the performance of the Pull-in Voltage performance in a MEMS switch. A hybrid model that combined DA with PSO was proposed in [88] to solve the optimal power flow (OPF) problem. DA was also used to select the optimal cluster heads in Radio Frequency Identification (RFID) networks [48]. Jafari and Chaleshtari [50] used the DA to optimize the orthotropic infinite plates with a quasi-triangular cut-out.

In [100], DA was utilized as an efficient optimizer for turn-mill operations. Hari-haran et al. [40] proposed an enhanced binary DA to deal with infant cry classification scenarios. Khadanga et al. [54] proposed a hybrid DA with pattern search for developing and studying of tilt integral derivative controllers. Kumar et al. [56] developed an improved fractional DA to be integrated with cluster cloud models. In [99], a modified cumulative DA in cooperation with neural networks was proposed. They also proposed a map-reduce cluster architecture to predict the performance of students using the DA-based approach. In [98], DA was employed to optimize the size and cost of static var compensator for enhancement of voltage profiles in power systems. Amroune et al. [15] proposed a new hybrid DA with support vector regression (SVR) for voltage stability analysis in power systems. Branch and Rey [21] proposed a DA-based load balancing approach to allocate resources in cloud computing. Babayigit [17] applied DA to synthesis of antenna arrays.

To sum up, in the reviewed papers, experimental results show the efficient performance of DA in terms of balancing the exploration and exploitation, speed of convergence, and satisfactory local optima (LO) avoidance.

5   Binary DA (BDA)

The DA algorithm was originally designed to solve the continuous optimization problems. Thus, using DA to tackle the FS problem requires changing the structure of the algorithm to deal with binary solutions. Transfer functions can be employed to convert the continuous solution to a binary one [77], by generating a probability of changing a position's elements to 0 or 1 based on the value of the step vector (velocity) of the ith search agent in the dth dimension in the current iteration (t) as an input parameter. Mirjalili [75] employed the transfer function in Eq. (8) to calculate the probability of changing the continuous positions to binary. This transfer function belongs the V-shaped (see Fig. 3) transfer functions category as denoted in [77].

$$T(v_d^i(t)) = \frac{|v_d^i(t)|}{\sqrt{1 + (v_d^i(t))^2}} \tag{8}$$

Fig. 3 V-Shaped transfer function [77]

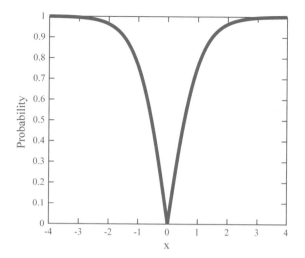

The result $T(v_k^i(t))$, obtained from Eq. (8) is then used to convert the i-th element of the position vector to 0 or 1 according to Eq. (9)

$$X(t+1) = \begin{cases} \neg X_t & r < T(v_k^i(t)) \\ X_t & r \geq T(v_k^i(t)) \end{cases} \tag{9}$$

where r is a random number in the [0,1] interval.

5.1 *BDA-Based Wrapper Feature Selection*

As the FS is a binary optimization problem, a binary vector of length N (where N is the number of features in the dataset), with 1 for the selected feature and 0 for the non selected feature, is used in the proposed approach (see Fig. 4). The KNN classifier [14] evaluates the selected feature subsets. Here, KNN classifier was used for its simplicity which makes it easy to implement, also it is a nonparametric classifier and hence can be used with different datasets. Moreover, KNN has been widely used with many FS methods and demonstrated to have competitive performance with a wide range of real life datasets [60].

In wrapper FS methods, two contradictory objectives should be considered; classification accuracy and reduction rate. The fitness function Eq. (10) considers both

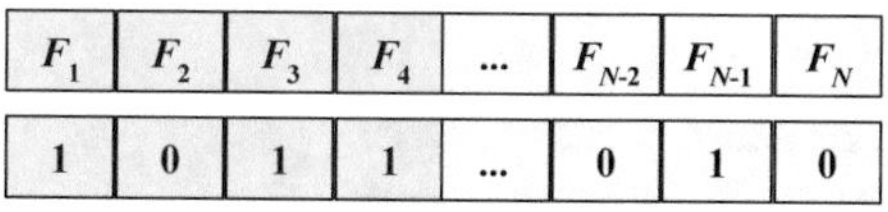

Fig. 4 Solution representation

F_1	F_2	F_3	F_4	...	F_{N-2}	F_{N-1}	F_N
1	0	1	1	...	0	1	0

aspects. α and β are two parameters corresponding to the importance of classification quality and subset length, α is in the [0,1] interval and $\beta =(1 -\alpha)$ is adapted from [69].

$$F(X) = \alpha * \gamma(X) + \beta * (1 - \frac{|R|}{|N|})$$ (10)

where $\gamma(X)$ represents the classification accuracy while using X feature subset, $|R|$ is the number of selected features and $|N|$ is the total number of features in the original dataset. Therefore, α (a number in the interval [0,1]) and β (1-α) [69] parameters were used to represent the importance of classification quality and the reduction rate, respectively.

Pseudocode of BDA-based feature selection technique is shown in Algorithm 2.

Algorithm 2 Pseudocode of BDA-based feature selection algorithm

Initialize the candidate features $X_i (i = 1, 2, \ldots, n)$
Initialize $\Delta X_i (i = 1, 2, \ldots, n)$
while (end condition is not met) **do**
    Evaluate all dragonflies based on Eq. (11.10)
    Update (F) and (E)
    Update coefficients $(i., e., w, s, a, c, f, and\ e)$
    Attain S, A, C, F, and E (based on Eqs. (9.21 to 4.5))
    Update step vectors (ΔX_{t+1}) by Eq. (4.6)
    Calculate $T(\Delta X)$ using Eq. (4.8)
    Update X_{t+1} by Eq. (4.9)
Return the best set of features

---

6 Results and Discussions

To test the proposed approach, 9 high dimensional datasets with low number of samples were used. As can be seen form Table 1, nine datasets have 226 distinct categories, 50308 samples (patients) and 2308–15009 variables (genes). All datasets are accessible in a public source.[1] A train/test model is utilized to validate the performance of the proposed BDA approach, where 80% of the data were used for training purposes, while the remaining 20% were used for testing [73]. All approaches were implemented using MATLAB 2017a and the results were repeated for 30 independent times on an Intel Core i5 machine, 2.2 GHz CPU and 4 GB of RAM. Table 2 presents the detailed parameter settings for utilized methods.

---

[1] http://www.gems-system.org/.

Table 1 Details of high-dimensional small instances datasets [92]

Dataset name	No. of samples	No. of features	No. of classes
11_Tumors	174	12533	11
14_Tumors	308	15009	26
Brain_Tumor1	90	5920	5
Brain_Tumor2	50	10367	4
Leukemia1	72	5327	3
Leukemia2	72	11225	3
SRBCT	83	2308	4
Prostate_Tumor	102	10509	2
DLBCL	77	5469	2

Table 2 The parameter settings

Parameter	Value
Population size	10
Number of iterations	100
Dimension	Number of features
Number of runs for each technique	30
α in fitness function	0.99
β in fitness function	0.01
a in bGWO	[2 0]
Q_{min} Frequency minimum in BBA	0
Q_{max} Frequency maximum in BBA	2
A Loudness in BBA	0.5
r Pulse rate in BBA	0.5
G_0 in BGSA	100
α in BGSA	20

6.1 Results and Analysis

In order to assess the performance of BDA algorithm on the high dimensional datasets, we tested six well-know metaheuristics algorithms for comparison purposes. All algorithms were implemented and ran on the same environment in order to make a fair comparison. For all algorithms, the average classification accuracy, average selected features, and average fitness values for the 30 independent runs are reported. Note that the best results in the subsequent results are highlighted in boldface. Moreover, a Wilcoxon signed-rank test is executed to state if there is a significant difference between the reported results with the significance interval 95% ($\alpha = 0.05$).

Table 3 shows the average classification accuracy over the 30 runs of different algorithms. As can be seen from Table 3, the proposed BDA shows the best performance compared to all other approaches. BGSA comes in the second place by obtaining the best results in three datasets, and then bGWO outperformed other approaches in only one dataset. Comparing BDA with BGSA; the second best approach, it can be obviously seen that BDA records the best results in 6 datasets, while BGSA is outperforming other approaches in 2 datasets, and both approaches obtained the same results in dealing with Brain_Tumor1 dataset. According to Table 4, which presents the p-values of BDA versus other approaches, there is a significant difference in the performance of the proposed approach and the other peers. These results verify that the proposed BDA maintains a more stable balance between diversification and intensification, and this leads to higher classification rates compared to other peers.

Table 3 Average classification accuracy for all approaches

Dataset	bGWO	BGSA	BBA	BGA	BPSO	BDA
11_Tumors	0.858	0.819	0.656	0.616	0.758	**0.879**
14_Tumors	0.576	0.481	0.556	0.428	0.477	**0.724**
Brain_Tumor1	**0.944**	**0.944**	0.894	0.848	0.726	0.889
Brain_Tumor2	0.720	0.800	0.543	0.745	0.416	**0.857**
DLBCL	0.754	**0.875**	0.848	0.868	0.818	**0.875**
Leukemia1	0.869	0.658	0.756	0.761	0.663	**0.933**
Leukemia2	0.867	0.867	0.922	0.821	0.818	**0.933**
Prostate_Tumor	0.878	**0.943**	0.921	0.834	0.806	0.902
SRBCT	0.882	0.884	0.859	0.760	0.746	**0.941**

Table 4 p-values between BDA and other approaches for the classification accuracy

Dataset	BDA versus				
	bGWO	BGSA	BBA	BGA	BPSO
11_Tumors	**1.61E-05**	**1.86E-11**	**1.20E-11**	**1.27E-11**	**1.19E-11**
14_Tumors	**2.54E-11**	**1.53E-11**	**2.64E-11**	**2.52E-11**	**2.84E-11**
Brain_Tumor1	**1.69E-14**	**1.69E-14**	1.70E-01	**7.24E-13**	**9.13E-13**
Brain_Tumor2	**1.03E-10**	**1.43E-06**	**8.50E-12**	**2.53E-11**	**8.71E-12**
DLBCL	**4.16E-14**	NaN	**2.98E-04**	**1.75E-02**	**9.21E-13**
Leukemia1	**1.17E-13**	**1.57E-13**	**3.37E-13**	**9.38E-13**	**1.02E-12**
Leukemia2	**1.69E-14**	**1.69E-14**	**2.14E-02**	**7.65E-13**	**5.57E-13**
Prostate_Tumor	**1.19E-04**	**2.98E-09**	**4.36E-03**	**9.49E-12**	**2.88E-12**
SRBCT	**1.69E-14**	**1.17E-13**	**3.48E-11**	**8.41E-13**	**8.09E-13**

Table 5 Average selected features for all approaches

Dataset	bGWO	BGSA	BBA	BGA	BPSO	BDA
11_Tumors	6233.23	6258.00	**5082.73**	6261.87	6261.50	5660.17
14_Tumors	8511.53	7564.97	**6207.97**	7500.77	7489.47	7206.77
Brain_Tumor1	2926.63	2896.13	2389.60	2960.17	2950.57	**2269.77**
Brain_Tumor2	5141.93	5106.47	**4173.83**	5156.87	5196.80	4285.23
DLBCL	2691.43	2699.57	2208.83	2731.20	2733.30	**1884.90**
Leukemia1	2620.63	2662.03	2077.73	2660.40	2664.53	**1850.47**
Leukemia2	5530.73	5523.27	4472.83	5613.60	5605.80	**4466.07**
Prostate_Tumor	5391.43	5242.90	**4124.50**	5269.43	5247.00	4372.77
SRBCT	1126.73	1144.80	925.37	1161.10	1148.97	**909.57**

Table 6 p-values between BDA and other approaches for the number of selected features

Dataset	BDA versus				
	bGWO	BGSA	BBA	BGA	BPSO
11_Tumors	**3.02E-11**	**3.02E-11**	**3.52E-07**	**3.01E-11**	**3.02E-11**
14_Tumors	**3.46E-10**	**1.46E-10**	**6.68E-11**	**2.22E-09**	**4.36E-09**
Brain_Tumor1	**3.02E-11**	**3.00E-11**	3.37E-01	**3.02E-11**	**3.01E-11**
Brain_Tumor2	**3.01E-11**	**3.01E-11**	1.02E-01	**3.01E-11**	**3.01E-11**
DLBCL	**3.01E-11**	**3.00E-11**	**2.60E-08**	**3.01E-11**	**3.02E-11**
Leukemia1	**3.00E-11**	**3.01E-11**	**7.20E-05**	**3.01E-11**	**3.01E-11**
Leukemia2	**3.01E-11**	**3.01E-11**	8.36E-01	**3.02E-11**	**3.01E-11**
Prostate_Tumor	**3.02E-11**	**3.01E-11**	**2.87E-02**	**3.01E-11**	**3.02E-11**
SRBCT	**3.00E-11**	**3.00E-11**	6.41E-01	**2.99E-11**	**3.00E-11**

Inspecting the results in Table 5, it can be obviously seen that BDA selects the minimal number of features on 55% of the datasets. It is followed by BBA which came in the second place by outperforming other approaches in 45% of the datasets. we see that no other approach is capable to compete with those two approaches. P-values in Table 6 also show that BDA can significantly outperform other approaches (namely bGWO, BGSA, BGA, and BPSO) in all datasets, while on five datasets out of nine, it can significantly outperform BBA approach.

As mentioned before, the adopted fitness function considers two objectives of the wrapper FS approach, i.e., the classification accuracy and the number of selected features. Observing the results in Tables 7 and 8, it can be seen the BDA significantly outperforms bGWO, BGSA, BGA, and BPSO approaches on all datasets, and on 55% of the datasets when compared to BBA. In addition to the superior performance of BDA approach, it recorded the fastest convergence rates among other approaches in most of the datasets as can be seen in Fig. 5, which shows the convergence speed of bGWO, BGSA, BBA, and BDA over the iterations. The reasons for satisfying

Table 7 Average fitness values for all approaches

Dataset	bGWO	BGSA	BBA	BGA	BPSO	BDA
11_Tumors	0.145	0.184	0.321	0.757	0.777	**0.124**
14_Tumors	0.425	0.519	0.417	0.473	0.504	**0.278**
Brain_Tumor1	0.060	**0.060**	0.080	0.837	0.876	0.114
Brain_Tumor2	0.282	0.203	0.393	0.600	0.653	**0.146**
DLBCL	0.248	0.129	**0.125**	0.843	0.845	0.127
Leukemia1	0.135	0.344	0.212	0.862	0.875	**0.069**
Leukemia2	0.137	0.137	**0.069**	0.863	0.884	0.070
Prostate_Tumor	0.126	0.062	**0.024**	0.846	0.859	0.102
SRBCT	0.121	0.119	0.105	0.864	0.889	**0.062**

Table 8 p-values between BDA and other approaches for the fitness values

Dataset	BDA versus				
	bGWO	BGSA	BBA	BGA	BPSO
11_Tumors	**7.67E-09**	**2.99E-11**	**3.01E-11**	**2.91E-11**	**2.94E-11**
14_Tumors	**3.01E-11**	**3.00E-11**	**3.01E-11**	**2.98E-11**	**2.96E-11**
Brain_Tumor1	**3.01E-11**	**3.00E-11**	**1.59E-07**	**2.82E-11**	**2.54E-11**
Brain_Tumor2	**2.94E-11**	**2.95E-11**	**3.01E-11**	**2.83E-11**	**2.68E-11**
DLBCL	**2.92E-11**	**2.95E-11**	1.20E-01	**2.57E-11**	**2.77E-11**
Leukemia1	**2.95E-11**	**3.01E-11**	**3.00E-11**	**2.76E-11**	**2.69E-11**
Leukemia2	**2.91E-11**	**2.91E-11**	**1.67E-06**	**2.80E-11**	**2.63E-11**
Prostate_Tumor	**1.28E-09**	**9.79E-05**	**4.97E-11**	**2.75E-11**	**2.80E-11**
SRBCT	**2.98E-11**	**2.96E-11**	**2.77E-05**	**2.79E-11**	**2.80E-11**

convergence curves in majority of datasets is that BDA has a high capacity for performing the focused intensification tendencies in last steps and extra diversification jumps in initial phases.

7 Conclusions and Future Directions

In this chapter, a wrapper FS approach (called BDA) that employs SSA as a selection strategy and KNN as an evaluator was proposed. BDA was tested using nine large scale medical datasets with low number of samples. Dealing with high dimensional datasets with low number of samples is challenging since the model cannot has no enough samples to be trained, in addition to having a large search space. BDA showed a good performance on those datasets when compared with five well known wrapper FS approaches.

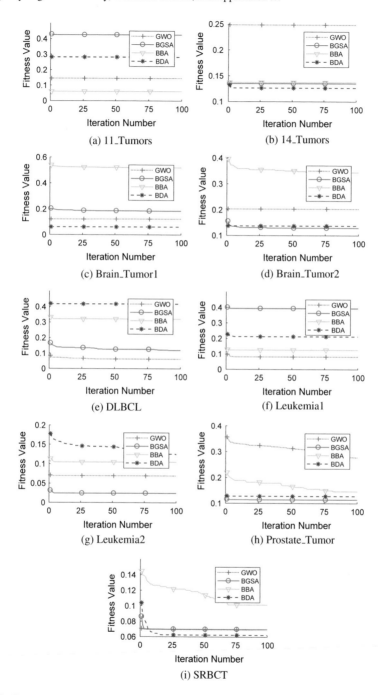

Fig. 5 Convergence curves for BGWO, BGSA, BBA, and BDA for all datasets

References

1. Abbassi, R., Abbassi, A., Heidari, A. A., & Mirjalili, S. (2019). An efficient salp swarm-inspired algorithm for parameters identification of photovoltaic cell models. *Energy Conversion and Management, 179*, 362–372.
2. An efficient binary salp swarm algorithm with crossover scheme for feature selection problems. (2018). *Knowledge-Based Systems, 154*, 43–67.
3. Abdel-Basset, M., Luo, Q., Miao, F., & Zhou, Y. (2017). Solving 0-1 knapsack problems by binary dragonfly algorithm. In *International Conference on Intelligent Computing* (pp. 491–502). Springer.
4. Al-Madi, N., Aljarah, I., & Ludwig, S. (2014). Parallel glowworm swarm optimization clustering algorithm based on mapreduce. In *IEEE Symposium Series on Computational Intelligence (IEEE SSCI 2014)*. IEEE Xplore Digital Library.
5. Aljarah, I., AlaM, A. Z., Faris, H., Hassonah, M. A., Mirjalili, S., & Saadeh, H. (2018). Simultaneous feature selection and support vector machine optimization using the grasshopper optimization algorithm. *Cognitive Computation*, 1–18.
6. Aljarah, I., Faris, H., & Mirjalili, S. (2018). Optimizing connection weights in neural networks using the whale optimization algorithm. *Soft Computing, 22*(1), 1–15.
7. Aljarah, I., Faris, H., Mirjalili, S., & Al-Madi, N. (2018). Training radial basis function networks using biogeography-based optimizer. *Neural Computing and Applications, 29*(7), 529–553.
8. Aljarah, I., & Ludwig, S. A. (2012). Parallel particle swarm optimization clustering algorithm based on mapreduce methodology. In *Proceedings of the Fourth World Congress on Nature and Biologically Inspired Computing (IEEE NaBIC12)*. IEEE Explore.
9. Aljarah, I., & Ludwig, S. A. (2013). A mapreduce based glowworm swarm optimization approach for multimodal functions. In *IEEE Symposium Series on Computational Intelligence, IEEE SSCI 2013*. IEEE Xplore.
10. Aljarah, I., & Ludwig, S. A. (2013). A new clustering approach based on glowworm swarm optimization. In *Proceedings of 2013 IEEE Congress on Evolutionary Computation Conference (IEEE CEC13)*. Cancun, Mexico: IEEE Xplore.
11. Aljarah, I., & Ludwig, S. A. (2013). Towards a scalable intrusion detection system based on parallel PSO clustering using mapreduce. In *Proceedings of Genetic and Evolutionary Computation Conference (ACM GECCO13)*. Amsterdam: ACM.
12. Aljarah, I., & Ludwig, S. A. (2016). A scalable mapreduce-enabled glowworm swarm optimization approach for high dimensional multimodal functions. *International Journal of Swarm Intelligence Research (IJSIR), 7*(1), 32–54.
13. Aljarah, I., Mafarja, M., Heidari, A. A., Faris, H., Zhang, Y., & Mirjalili, S. (2018). Asynchronous accelerating multi-leader salp chains for feature selection. *Applied Soft Computing, 71*, 964–979.
14. Altman, N. S. (1992). An introduction to kernel and nearest-neighbor nonparametric regression. *The American Statistician, 46*(3), 175–185.
15. Amroune, M., Bouktir, T., & Musirin, I. (2018). Power system voltage stability assessment using a hybrid approach combining dragonfly optimization algorithm and support vector regression. *Arabian Journal for Science and Engineering*, 1–14.
16. Aminisharifabad, M., Yang, Q., & Wu, X. (2018). A penalized autologistic regression with application for modeling the microstructure of dual-phase high strength steel. *Journal of Quality Technology,* in-press.
17. Babayigit, B. (2018). Synthesis of concentric circular antenna arrays using dragonfly algorithm. *International Journal of Electronics, 105*(5), 784–793.
18. Barham, R., & Aljarah, I. (2017). Link prediction based on whale optimization algorithm. In *The International Conference on new Trends in Computing Sciences (ICTCS2017)*. Amman: Jordan.

19. Bhesdadiya, R., Pandya, M. H., Trivedi, I. N., Jangir, N., Jangir, P., & Kumar, A. (2016). Price penalty factors based approach for combined economic emission dispatch problem solution using dragonfly algorithm. In *2016 International Conference on Energy Efficient Technologies for Sustainability (ICEETS)* (pp. 436–441). IEEE.
20. Blum, A. L., & Langley, P. (1997). Selection of relevant features and examples in machine learning. *Artificial Intelligence, 97*(1–2), 245–271.
21. Branch, S. R., & Rey, S. (2018). Providing a load balancing method based on dragonfly optimization algorithm for resource allocation in cloud computing. *International Journal of Networked and Distributed Computing, 6*(1), 35–42.
22. Chen, Y., Li, L., Xiao, J., Yang, Y., Liang, J., & Li, T. (2018). Particle swarm optimizer with crossover operation. *Engineering Applications of Artificial Intelligence, 70*, 159–169.
23. Chitsaz, H., & Aminisharifabad, M. (2015). Exact learning of rna energy parameters from structure. *Journal of Computational Biology, 22*(6), 463–473.
24. Daely, P. T., & Shin, S. Y. (2016). Range based wireless node localization using dragonfly algorithm. In *2016 Eighth International Conference on Ubiquitous and Future Networks (ICUFN)* (pp. 1012–1015). IEEE.
25. Dorigo, M., & Birattari, M. (2011). Ant colony optimization. In *Encyclopedia of machine learning* (pp. 36–39). Springer.
26. Dorigo, M., & Di Caro, G. (1999). Ant colony optimization: a new meta-heuristic. In *Proceedings of the 1999 Congress on Evolutionary Computation, 1999. CEC 99*, vol. 2 (pp. 1470–1477). IEEE.
27. Eberhart, R., & Kennedy, J. (1995). A new optimizer using particle swarm theory. In *Proceedings of the Sixth International Symposium on Micro Machine and Human Science, 1995. MHS'95* (pp. 39–43). IEEE.
28. Elhariri, E., El-Bendary, N., & Hassanien, A. E. (2016). Bio-inspired optimization for feature set dimensionality reduction. In *2016 3rd International Conference on Advances in Computational Tools for Engineering Applications (ACTEA)* (pp. 184–189). IEEE.
29. Faris, H., Aljarah, I., Al-Madi, N., & Mirjalili, S. (2016). Optimizing the learning process of feedforward neural networks using lightning search algorithm. *International Journal on Artificial Intelligence Tools, 25*(06), 1650033.
30. Faris, H., Aljarah, I., & Al-Shboul, B. (2016). A hybrid approach based on particle swarm optimization and random forests for e-mail spam filtering. *International Conference on Computational Collective Intelligence* (pp. 498–508). Cham: Springer.
31. Faris, H., Aljarah, I., & Mirjalili, S. (2017). Evolving radial basis function networks using moth–flame optimizer. In *Handbook of Neural Computation* (pp. 537–550).
32. Faris, H., Aljarah, I., et al. (2015). Optimizing feedforward neural networks using krill herd algorithm for e-mail spam detection. In *2015 IEEE Jordan Conference on Applied Electrical Engineering and Computing Technologies (AEECT)* (pp. 1–5). IEEE.
33. Faris, H., Ala'M, A. Z., Heidari, A. A., Aljarah, I., Mafarja, M., Hassonah, M. A., & Fujita, H. (2019). An intelligent system for spam detection and identification of the most relevant features based on evolutionary random weight networks. *Information Fusion, 48*, 67–83.
34. Faris, H., Mafarja, M. M., Heidari, A. A., Aljarah, I., AlaM, A. Z., Mirjalili, S., et al. (2018). An efficient binary salp swarm algorithm with crossover scheme for feature selection problems. *Knowledge-Based Systems, 154*, 43–67.
35. Fisher, L. (2009). *The perfect swarm: The science of complexity in everyday life*. Basic Books.
36. Ghatasheh, N., Faris, H., Aljarah, I., & Al-Sayyed, R. M. (2015). Optimizing software effort estimation models using firefly algorithm. *Journal of Software Engineering and Applications, 8*(03), 133.
37. Guha, K., Laskar, N., Gogoi, H., Borah, A., Baishnab, K., & Baishya, S. (2017). Novel analytical model for optimizing the pull-in voltage in a flexured mems switch incorporating beam perforation effect. *Solid-State Electronics, 137*, 85–94.
38. Guo, S., Dooner, M., Wang, J., Xu, H., & Lu, G. (2017). Adaptive engine optimisation using NSGA-II and MODA based on a sub-structured artificial neural network. In *2017 23rd International Conference on Automation and Computing (ICAC)* (pp. 1–6). IEEE.

39. Hamdy, M., Nguyen, A. T., & Hensen, J. L. (2016). A performance comparison of multi-objective optimization algorithms for solving nearly-zero-energy-building design problems. *Energy and Buildings, 121*, 57–71.
40. Hariharan, M., Sindhu, R., Vijean, V., Yazid, H., Nadarajaw, T., Yaacob, S., et al. (2018). Improved binary dragonfly optimization algorithm and wavelet packet based non-linear features for infant cry classification. *Computer Methods and Programs in Biomedicine, 155*, 39–51.
41. Heidari, A. A., & Abbaspour, R. A. (2018). Enhanced chaotic grey wolf optimizer for real-world optimization problems: A comparative study. In *Handbook of Research on Emergent Applications of Optimization Algorithms* (pp. 693–727). IGI Global.
42. Heidari, A. A., Abbaspour, R. A., & Jordehi, A. R. (2017). An efficient chaotic water cycle algorithm for optimization tasks. *Neural Computing and Applications, 28*(1), 57–85.
43. Heidari, A. A., Abbaspour, R. A., & Jordehi, A. R. (2017). Gaussian bare-bones water cycle algorithm for optimal reactive power dispatch in electrical power systems. *Applied Soft Computing, 57*, 657–671.
44. Heidari, A. A., & Delavar, M. R. (2016). A modified genetic algorithm for finding fuzzy shortest paths in uncertain networks. In *ISPRS - International Archives of the Photogrammetry, Remote Sensing and Spatial Information Sciences XLI-B2* (pp. 299–304).
45. Heidari, A. A., Faris, H., Aljarah, I., & Mirjalili, S. (2018). An efficient hybrid multilayer perceptron neural network with grasshopper optimization. *Soft Computing*, 1–18.
46. Heidari, A. A., & Pahlavani, P. (2017). An efficient modified grey wolf optimizer with lévy flight for optimization tasks. *Applied Soft Computing, 60*, 115–134.
47. Hema, C., Sankar, S., et al. (2016). Energy efficient cluster based protocol to extend the RFID network lifetime using dragonfly algorithm. In *2016 International Conference on Communication and Signal Processing (ICCSP)* (pp. 0530–0534). IEEE.
48. Hema, C., Sankar, S., et al. (2017). Performance comparison of dragonfly and firefly algorithm in the RFID network to improve the data transmission. *Journal of Theoretical and Applied Information Technology, 95*(1), 59.
49. Ibrahim, H. T., Mazher, W. J., Ucan, O. N., & Bayat, O. (2018). A grasshopper optimizer approach for feature selection and optimizing SVM parameters utilizing real biomedical data sets. *Neural Computing and Applications*.
50. Jafari, M., & Chaleshtari, M. H. B. (2017). Using dragonfly algorithm for optimization of orthotropic infinite plates with a quasi-triangular cut-out. *European Journal of Mechanics-A/Solids, 66*, 1–14.
51. Karaboga, D., & Basturk, B. (2007). A powerful and efficient algorithm for numerical function optimization: Artificial bee colony (ABC) algorithm. *Journal of Global Optimization, 39*(3), 459–471.
52. Kashef, S., & Nezamabadi-pour, H. (2015). *An advanced ACO algorithm for feature subset selection, 147*, 271–279.
53. Kennedy, J. (2006). Swarm intelligence. In *Handbook of Nature-Inspired and Innovative Computing* (pp. 187–219). Springer.
54. Khadanga, R. K., Padhy, S., Panda, S., & Kumar, A. (2018). Design and analysis of tilt integral derivative controller for frequency control in an islanded microgrid: A novel hybrid dragonfly and pattern search algorithm approach. *Arabian Journal for Science and Engineering*, 1–12.
55. Ks, S. R., & Murugan, S. (2017). Memory based hybrid dragonfly algorithm for numerical optimization problems. *Expert Systems with Applications, 83*, 63–78.
56. Kumar, C. A., Vimala, R., Britto, K. A., & Devi, S. S. (2018). FDLA: Fractional dragonfly based load balancing algorithm in cluster cloud model. *Cluster Computing*, 1–14.
57. Langley, P., et al. (1994). Selection of relevant features in machine learning. *Proceedings of the AAAI Fall symposium on relevance, 184*, 245–271.
58. Li, Q., Chen, H., Huang, H., Zhao, X., Cai, Z., Tong, C., et al. (2017). An enhanced grey wolf optimization based feature selection wrapped kernel extreme learning machine for medical diagnosis. *Computational and Mathematical Methods in Medicine, 2017*.

59. Li, Y., Li, T., & Liu, H. (2017). Recent advances in feature selection and its applications. *Knowledge and Information Systems, 53*(3), 551–577.
60. Liao, T., & Kuo, R. (2018). Five discrete symbiotic organisms search algorithms for simultaneous optimization of feature subset and neighborhood size of KNN classification models. *Applied Soft Computing, 64*, 581–595.
61. Liu, H., & Motoda, H. (2012). *Feature selection for knowledge discovery and data mining,* vol. 454. Springer Science & Business Media.
62. Liu, H., Setiono, R., et al. (1996). A probabilistic approach to feature selection-a filter solution. In *Thirteenth International Conference on Machine Learning (ICML)*, vol. 96 (pp. 319–327). Citeseer.
63. Mafarja, M., & Abdullah, S. (2011). Modified great deluge for attribute reduction in rough set theory. In *2011 Eighth International Conference on Fuzzy Systems and Knowledge Discovery (FSKD)*, vol. 3, (pp. 1464–1469). IEEE.
64. Mafarja, M., & Abdullah, S. (2013). Investigating memetic algorithm in solving rough set attribute reduction. *International Journal of Computer Applications in Technology, 48*(3), 195–202.
65. Mafarja, M., & Abdullah, S. (2013). Record-to-record travel algorithm for attribute reduction in rough set theory. *Journal of Theoretical and Applied Information Technology, 49*(2), 507–513.
66. Mafarja, M., & Abdullah, S. (2014). Fuzzy modified great deluge algorithm for attribute reduction. *Recent Advances on Soft Computing and Data Mining* (pp. 195–203). Cham: Springer.
67. Mafarja, M., & Abdullah, S. (2015). A fuzzy record-to-record travel algorithm for solving rough set attribute reduction. *International Journal of Systems Science, 46*(3), 503–512.
68. Mafarja, M., Aljarah, I., Heidari, A. A., Faris, H., Fournier-Viger, P., Li, X., & Mirjalili, S. (2018). Binary dragonfly optimization for feature selection using time-varying transfer functions. *Knowledge-Based Systems, 161*, 185–204.
69. Mafarja, M., Aljarah, I., Heidari, A. A., Hammouri, A. I., Faris, H., AlaM, A. Z., et al. (2018). Evolutionary population dynamics and grasshopper optimization approaches for feature selection problems. *Knowledge-Based Systems, 145*, 25–45.
70. Mafarja, M., Eleyan, D., Abdullah, S., & Mirjalili, S. (2017). S-shaped vs. V-shaped transfer functions for ant lion optimization algorithm in feature selection problem. In *Proceedings of the International Conference on Future Networks and Distributed Systems* (p. 1). ACM.
71. Mafarja, M., Jaber, I., Eleyan, D., Hammouri, A., & Mirjalili, S. (2017). Binary dragonfly algorithm for feature selection. In *2017 International Conference on New Trends in Computing Sciences (ICTCS)* (pp. 12–17).
72. Mafarja, M., & Mirjalili, S. (2017). Hybrid whale optimization algorithm with simulated annealing for feature selection. *Neurocomputing*.
73. Mafarja, M., & Mirjalili, S. (2017). Whale optimization approaches for wrapper feature selection. *Applied Soft Computing, 62*, 441–453.
74. Mirjalili, S. (2015). The ant lion optimizer. *Advances in Engineering Software, 83*, 80–98.
75. Mirjalili, S. (2016). Dragonfly algorithm: A new meta-heuristic optimization technique for solving single-objective, discrete, and multi-objective problems. *Neural Computing and Applications, 27*(4), 1053–1073.
76. Mirjalili, S., Gandomi, A. H., Mirjalili, S. Z., Saremi, S., Faris, H., & Mirjalili, S. M. (2017). Salp swarm algorithm: A bio-inspired optimizer for engineering design problems. *Advances in Engineering Software, 114*, 163–191.
77. Mirjalili, S., & Lewis, A. (2013). S-shaped versus V-shaped transfer functions for binary particle swarm optimization. *Swarm and Evolutionary Computation, 9*, 1–14.
78. Mirjalili, S., & Lewis, A. (2016). The whale optimization algorithm. *Advances in Engineering Software, 95*, 51–67.
79. Mirjalili, S., Mirjalili, S. M., & Lewis, A. (2014). Grey wolf optimizer. *Advances in Engineering Software, 69*, 46–61.
80. Moradi, P., & Gholampour, M. (2016). A hybrid particle swarm optimization for feature subset selection by integrating a novel local search strategy. *Applied Soft Computing, 43*, 117–130.

81. Pathania, A. K., Mehta, S., & Rza, C. (2016). Economic load dispatch of wind thermal integrated system using dragonfly algorithm. In *2016 7th India International Conference on Power Electronics (IICPE)* (pp. 1–6). IEEE.
82. Rajamohana, S., & Umamaheswari, K. (2018). Hybrid approach of improved binary particle swarm optimization and shuffled frog leaping for feature selection. *Computers & Electrical Engineering*.
83. Raman, G., Raman, G., Manickam, C., & Ganesan, S. I. (2016). Dragonfly algorithm based global maximum power point tracker for photovoltaic systems. In Y. Tan, Y. Shi, & B. Niu (Eds.), *Advances in Swarm Intelligence* (pp. 211–219). Cham: Springer International Publishing.
84. Reynolds, C. W. (1987). Flocks, herds and schools: A distributed behavioral model. In *ACM SIGGRAPH Computer Graphics*, vol. 21 (pp. 25–34). ACM.
85. Salam, M. A., Zawbaa, H. M., Emary, E., Ghany, K. K. A., & Parv, B. (2016). A hybrid dragonfly algorithm with extreme learning machine for prediction. In *2016 International Symposium on INnovations in Intelligent SysTems and Applications (INISTA)* (pp. 1–6). IEEE.
86. Sambandam, R. K., & Jayaraman, S. (2016). Self-adaptive dragonfly based optimal thresholding for multilevel segmentation of digital images. *Journal of King Saud University-Computer and Information Sciences*.
87. Saremi, S., Mirjalili, S., & Lewis, A. (2017). Grasshopper optimisation algorithm: Theory and application. *Advances in Engineering Software*, *105*, 30–47.
88. Shilaja, C., & Ravi, K. (2017). Optimal power flow using hybrid DA-APSO algorithm in renewable energy resources. *Energy Procedia*, *117*, 1085–1092.
89. Sekhar, A. H., & Devi, A. L. (2016). Analysis of multi tcsc placement in transmission system by using firing angle control model with heuristic algorithms. *ARPN Journal of Engineering and Applied Sciences*, *11*(21), 12743–12755.
90. Shukri, S., Faris, H., Aljarah, I., Mirjalili, S., & Abraham, A. (2018). Evolutionary static and dynamic clustering algorithms based on multi-verse optimizer. *Engineering Applications of Artificial Intelligence*, *72*, 54–66.
91. Song, J., & Li, S. (2017). Elite opposition learning and exponential function steps-based dragonfly algorithm for global optimization. In *2017 IEEE International Conference on Information and Automation (ICIA)* (pp. 1178–1183). IEEE.
92. Statnikov, A., Aliferis, C. F., Tsamardinos, I., Hardin, D., & Levy, S. (2004). A comprehensive evaluation of multicategory classification methods for microarray gene expression cancer diagnosis. *Bioinformatics*, *21*(5), 631–643.
93. Sugave, S. R., Patil, S. H., & Reddy, B. E. (2017). DDF: Diversity dragonfly algorithm for cost-aware test suite minimization approach for software testing. In *2017 International Conference on Intelligent Computing and Control Systems (ICICCS)* (pp. 701–707). IEEE.
94. Suresh, V., & Sreejith, S. (2017). Generation dispatch of combined solar thermal systems using dragonfly algorithm. *Computing*, *99*(1), 59–80.
95. Surowiecki, J., Silverman, M. P., et al. (2007). The wisdom of crowds. *American Journal of Physics*, *75*(2), 190–192.
96. Tharwat, A., Gabel, T., & Hassanien, A.E. (2017). Classification of toxicity effects of biotransformed hepatic drugs using optimized support vector machine. In A. E. Hassanien, K. Shaalan, T. Gaber, M. F. Tolba (Eds.), *Proceedings of the International Conference on Advanced Intelligent Systems and Informatics 2017* (pp. 161–170). Cham: Springer International Publishing.
97. Tharwat, A., Gabel, T., & Hassanien, A.E. (2017). Parameter optimization of support vector machine using dragonfly algorithm. In *International Conference on Advanced Intelligent Systems and Informatics* (pp. 309–319). Springer.
98. Vanishree, J., & Ramesh, V. (2018). Optimization of size and cost of static var compensator using dragonfly algorithm for voltage profile improvement in power transmission systems. *International Journal of Renewable Energy Research (IJRER)*, *8*(1), 56–66.
99. VeeraManickam, M., Mohanapriya, M., Pandey, B. K., Akhade, S., Kale, S., Patil, R., et al. (2018). Map-reduce framework based cluster architecture for academic students performance prediction using cumulative dragonfly based neural network. *Cluster Computing*, 1–17.

100. Vikram, K. A., Ratnam, C., Lakshmi, V., Kumar, A. S., & Ramakanth, R. (2018). Application of dragonfly algorithm for optimal performance analysis of process parameters in turn-mill operations-a case study. In *IOP Conference Series: Materials Science and Engineering*, vol. 310 (p. 012154). IOP Publishing.
101. Wu, J., Zhu, Y., Wang, Z., Song, Z., Liu, X., Wang, W., et al. (2017). A novel ship classification approach for high resolution sar images based on the BDA-KELM classification model. *International Journal of Remote Sensing, 38*(23), 6457–6476.
102. Zorarpacı, E., & Özel, S. A. (2016). A hybrid approach of differential evolution and artificial bee colony for feature selection. *Expert Systems with Applications, 62*, 91–103.

Genetic Algorithm: Theory, Literature Review, and Application in Image Reconstruction

Seyedali Mirjalili, Jin Song Dong, Ali Safa Sadiq and Hossam Faris

Abstract Genetic Algorithm (GA) is one of the most well-regarded evolutionary algorithms in the history. This algorithm mimics Darwinian theory of survival of the fittest in nature. This chapter presents the most fundamental concepts, operators, and mathematical models of this algorithm. The most popular improvements in the main component of this algorithm (selection, crossover, and mutation) are given too. The chapter also investigates the application of this technique in the field of image processing. In fact, the GA algorithm is employed to reconstruct a binary image from a completely random image.

1 Introduction

One of the fast growing sub-fields of Computational Intelligence is Evolutionary Computation. In this field, there are a large number of algorithms to solve optimization problems. Such algorithms mostly mimic biological evolutionary in nature. The

S. Mirjalili (✉) · J. Song Dong
Institute for Integrated and Intelligent Systems, Griffith University, Nathan,
Brisbane, QLD 4111, Australia
e-mail: seyedali.mirjalili@griffithuni.edu.au

J. Song Dong
Department of Computer Science, School of Computing, National University of Singapore,
Singapore, Singapore
e-mail: j.dong@griffith.edu.au

A. S. Sadiq
School of Information Technology, Monash University, 47500 Bandar Sunway, Malaysia
e-mail: ali.safaa@monash.edu

H. Faris
King Abdullah II School for Information Technology, The University of Jordan,
Amman, Jordan
e-mail: hossam.faris@ju.edu.jo

© Springer Nature Switzerland AG 2020
S. Mirjalili et al. (eds.), *Nature-Inspired Optimizers*, Studies in Computational
Intelligence 811, https://doi.org/10.1007/978-3-030-12127-3_5

framework of most of Evolutionary Algorithms (EAs) are very similar. They start with a population of random solutions. Each solution is evaluated by a fitness function that indicates the suitability of the solutions. Through several iterations, the best solutions are chosen. The best solutions together with stochastic selections are combined to produce the next set of solutions. Evolutionary algorithms are equipped with several random (stochastic) components which select and combine solutions in each population. This makes them unreliable in finding similar solution in each run as opposed to deterministic algorithms. Deterministic algorithms (e.g. brute force search) find the same solution in every run. However, they suffer from slower speed and local solution stagnation when applied to large-scale problems.

EAs are stochastic and mostly heuristics. This means to search only a part of search space using heuristics information. Finding the best solutions in each population and using them to improve other solutions allow theses algorithms to search promising regions of the search space instead of all. The wide range of large-scale applications of EAs in recent years show the popularity and flexibility of such techniques. Another advantage of EAs is that they consider optimization problems as block boxes. The shape of search space and its derivation should not be known to users of EAs as opposed to gradient-based optimization algorithms. One of the first and most well-regarded EAs is Genetic Algorithm (GA). This algorithm is discussed and analyzed in this chapter.

2 Genetic Algorithms

2.1 Inspiration

The GA algorithm is inspired by the theory of biological evolution proposed by Darwin [1, 2]. In fact, the main mechanism simulated in GA is the survival of the fittest. In nature, fitter organisms have a higher probability of survival. This assists them to transfer their genes to the next generation. Over time, good genes that allow species to better adapt with environment (avoid enemy and find food) become dominant in subsequent generations.

Inspired by chromosomes and genes in nature, the GA algorithm represents optimization problems as a set of variables. Each solution for an optimization problem corresponds to a chromosome and each gene corresponds to a variable of the problem. For instance, for a problem with 10 variables, GA uses chromosomes with 10 genes. The GA algorithm uses three main operators to improve the chromosomes in each generation: selection, crossover (recombination), and mutation. These steps as well as problem representation and initial population are discussed in the following sub-sections.

2.2 Gene Representation

As discussed above, each chromosome corresponds to a candidate solution to a given optimization problem. A chromosome is made of multiple genes that simulate the variables of the optimization problem. To use the GA algorithm, therefore, the first step is to formulate the problem and define its parameters as a vector. There are two variants of GA: binary and continuous. In the binary version, each gene can be assigned with two values (e.g. 0 or 1). In the continuous version, any continuous values with a set of upper and lower bounds are used. A special case of the binary GA is where there are more than two values to choose from. In this case, more bits should be allocated to the variables of the problem. For instance, if the problem has two variables each of which can be assigned with eight different values, we need three genes for each of the variables. Therefore, the number of genes to selection n discrete values is calculated as $log_2 n$.

It should be noted that genes can be even characters or components of a program. As long as the genes are fed into a fitness function and lead to a fitness value, they can be used in the GA algorithm. In the case of using different parts of a computer programs for each gene, GA is called Genetic Programming. In this chapter, binary and continuous GA are discussed and used.

2.3 Initial Population

GA starts the optimization process with a set of random chromosomes. In case of binary GA, the following code is used:

$$X_i = \begin{cases} 1 & r_i < 0.5 \\ 0 & otherwise \end{cases} \tag{1}$$

where X_i is the i-th gene and r_i is a random number in [0,1] generated separately for each gene.

In the continuous GA, the following equation is used to randomly initialize the genes:

$$X_i = (ub_i - lb_i) * r_i + lb_i \tag{2}$$

where X_i is the i-th gene, r_i is a random number in [0,1] generated separately for each gene, ub_i is the upper bound for the i-th gene (variable), and lb_i is the lower bound for the i-th gene(variable).

The main objective in this phase is to have uniformly distributed random solutions across all of the variables since these solutions will be used in the following operators.

2.4   Selection

The selection operator of GA simulates the natural selection. In natural selection, the chance of survival is increased proportional to the fitness. Being selected leads to the propagation of their genes to subsequent generations. In GA, this is can be simulated with a roulette wheel. Figure 1 illustrates an example of a roulette wheel for six individuals. The details of these individuals are presented in Table 1. The figure and the table show that the individuals with higher fitness values have a higher chance of being selected by the roulette wheel.

The roulette wheel normalizes and maps the fitness values to probability values. This means that the lower and upper bounds of the roulette wheel are 0 and 1 respectively. By generating a random number in this interval, one of the individuals will be selected. The larger sector in the pi chart of the roulette wheel that an individual occupies, a higher chance that the individual being selected using this operator.

An important question here is why poor individuals are not discarded. In nature, less fit individual might also be able to mate and contribute to the production of subsequent generations. This depends on environmental situations, territory, and competition. A poor individual might have genes that produce an excellent feature in conjunction with genes of another individual to combat a new change in the environment. Therefore, giving a small chance of selection to poor individuals keeps a small probability of taking advantages of their good features.

Note that the range of fitness values changes in different problems, so such values should be normalized as can be illustrated in Table 1. Also, one of the issues of using the roulette wheel is that it fails in handling negative fitness values. Due to calculating

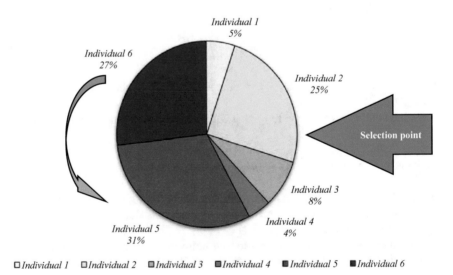

Fig. 1 Mechanism of the roulette wheel in GA. The best individual (#5) has the largest share of the roulette wheel, while the worst individual (#4) has the lowest share

Table 1 Details of the individuals in Fig. 1. The fittest individual is Individual #5

Individual number	Fitness value	% of Total
1	12	5
2	55	24
3	20	8
4	10	4
5	70	30
6	60	26
Total	227	100

cumulative sum during the process of simulating the roulette wheel, fitness scaling should be done to change all the negative fitness values to positive ones.

It should be noted that the roulette wheel is one of the many selection operators in the literature [3–5]. Some of other selection operators are:

- Boltzmann selection [6]
- Tournament selection [7]
- Rank selection [8]
- Steady state selection [9]
- Truncation selection [10]
- Local selection [11]
- Fuzzy selection [12]
- Fitness uniform selection [13]
- Proportional selection [14]
- Linear rank selection [14]
- Steady-state reproduction [15]

2.5 Crossover (Recombination)

The natural selection presented in the previous subsection allows selecting individuals as parents for the crossover step. This step allows exchanging genes between individuals to produce new solutions. There are different methods of crossover in the literature. The easiest one divides chromosomes into two (single-point) or three pieces (double-point) [16]. They then exchange genes between two chromosomes. An example is visualized in Fig. 2.

In the single-point cross over, the chromosomes of two parent solutions are swapped before and after a single point. In the double-point crossover, however, there are two cross over points and the chromosomes between the points are swapped only. Other crossover techniques in the literature are:

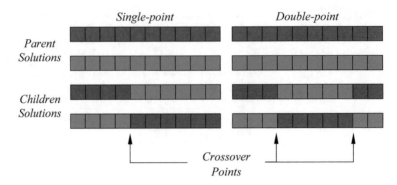

Fig. 2 Two popular crossover techniques in GA: single-point and double point. In the single-point cross over, the chromosomes of two parent solutions are swapped before and after a single point. In the double-point crossover, there are two cross over points and the chromosomes between the points are swapped only

- Uniform crossover [17]
- Half uniform crossover [18]
- Three parents crossover [19]
- Partially matched crossover [20]
- Cycle crossover [21]
- Order crossover [22]
- Position-based crossover [23]
- Heuristic cross over [24]
- Masked crossover [25]
- Multi-point crossover [26]

The main objective of crossover is to make sure that the genes are exchanged and the children inherit the genes from the parents. Crossover is the main mechanism of exploitation in GA. For two given parents, if we assume that the crossover is done using random crossover point, the algorithm is trying to check and search different combinations of genes coming from the parents. This leads to exploitation of those possible solutions without introducing even a single new gene.

Note that there is a parameter in GA called Probability of Crossover (P_c) which indicates the probability of accepting a new child. This parameter is a number in the interval of [0,1]. A random number in the same interval is generated for each child. If this random number is less than P_c, the child is propagated to the subsequent generation. Otherwise, the parent will be propagated. This happens in nature as well—all of the offspring do not survive.

2.6 *Mutation*

In the previous subsection, it was discussed that the crossover operator exchanges genes between chromosomes. The issue with this mechanism is the lack of introducing new genes. If all the solutions become poor (be trapped in locally optimal solutions), crossover does not lead to different solutions with new genes differing from those in the parents. To address this, the mutation operator is also considered in the GA algorithm.

Mutation causes random changes in the genes. There is a parameter called Probability of Mutation (P_m) that is used for every gene in a child chromosome produced in the crossover stage. This parameter is a number in the interval of [0,1]. A random number in the same interval is generated for each gene in the new child. If this random number is less than P_m, the gene is assigned with a random number with the lower and upper bounds. The impact of the mutation operator on the children chromosomes are shown in Fig. 3. It can be seen in this figure that slight changes in some of the randomly selected genes occur after the crossover (recombination) phase.

There are many mutation techniques in the literate too as follows:

- Power mutation [27]
- Uniform [28]
- Non-uniform [29]
- Gaussian [30]
- Shrink [31]
- Supervised mutation [32]
- Uniqueness mutation [33]
- Varying probability of mutation [34]

It is worth discussing here that the mutation is the main mechanism of exploration in GA. This is because this operator introduces random changes and causes a solution to go beyond the current region of the search space to regions that are not associated with the genes inherited from the parents.

Fig. 3 Mutation operator alters one or multiple genes in the children solutions after the crossover phase

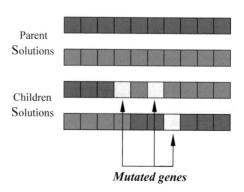

2.7   Elitism

Crossover and mutation change the genes in the chromosomes. Depending on the P_m, there is a chance that all the parents are replaced by the children. This might lead to the problem of losing good solutions in the current generation. To fix this, we can use a new operator called Elitism [35]. This operator was not initially integrated into the first version of GA. However, a large number of researchers used it recently due to its significant impact on the performance of GA.

The mechanism of this operator is very simple. A portion of the best chromosomes in the current population is maintained and propagated to the subsequent generation without any changes. This prevents those solutions from being damaged by the crossover and mutation during the process of creating the new populations. The list of elites gets updated by simply raking the individuals based on their fitness values.

2.8   Continuous GA

The GA algorithm presented so far is able to solve binary problems. Based on this algorithm, there are different methods to solve continuous problems. One of the easiest methods only modifies the crossover operator using the following equations:

$$Child1_i = \beta_i * Parent1_i + (1 - \beta_i) * Parent2_i \qquad (3)$$
$$Child2_i = (1 - \beta_i) * Parent_i + \beta_i * Parent2_i \qquad (4)$$

where β_i is a random number generated for the i-th gene in the chromosome.

These two equations use the addition numerical operator that generates continuous numbers. So, the genes can be assigned with any numbers in an interval.

3   Experiments

Image reconstruction [36] is a popular method in the field of image processing. There are several applications for this technique including noise reduction in satellite images or reconstruction of corrupted images. In this section the performance of GA is investigated in constructing binary images. A number of binary images with different dimensions are employed as case studies. The GA algorithm starts with a randomly constructed image. It then tries to minimize the discrepancy of the construction images with the actual binary image as the ground truth.

GA with 50 chromosomes and maximum of 2000 generations is employed. The probability crossover is 0.95, the probability of mutation is 0.001, and the elitism rate is 20%. The stopping condition is either reaching the maximum number of iterations or 0 in error (the discrepancy). The GA employed uses a roulette when in the selection and double-point crossover.

The binary images are divided into three classes: images with simple patterns, images with complicated patterns, and images of different objects. Their sizes are also increased. In each chromosome, there are $r \times c$ genes, which is the total number of binary pixels in a binary image with r rows and c columns.

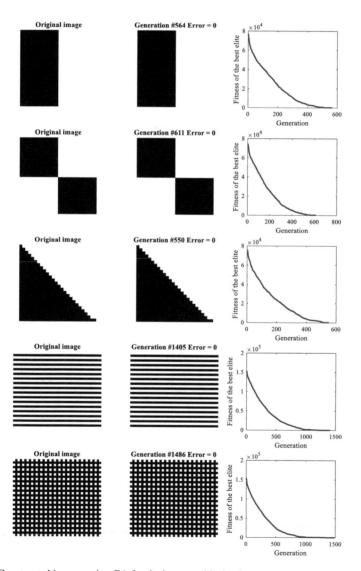

Fig. 4 Constructed images using GA for the images with simple patterns

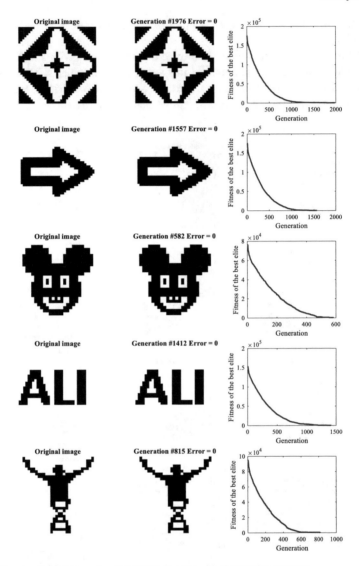

Fig. 5 Constructed images using GA for the images with challenging patterns

Fig. 6 Constructed images
using GA for the image with
a very challenging pattern

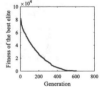

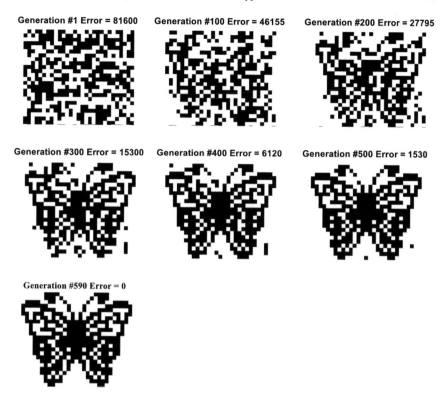

Fig. 7 Improvements (evolving) in the constructed image over 2000 generations

Figure 4 shows the results of GA on case studies with simple patterns. Note that the number of pixels to be optimized in these figures are equal to 650 (26*25). It can be seen that in all cases the GA algorithm finds the best solution less than the maximum number of generations. The convergence curves are also constantly improved, showing a consistent exploratory and exploitative behavior of GA. This is one of the advantages of GA, in which there is no adaptive mechanism. The probability of crossover and mutation do not change, but the algorithm maintains the best solutions (elite) and uses their good chromosomes to improve the quality of the entire population.

Figure 5 shows that GA performs well on complex shapes as well when keeping the same image size. This figure shows that the GA finds an identical approximation of the image in less than 2000 generations. This shows that the pattern does not matter here as long as the size of the image does not change, the same configuration can find the best solution. Another challenging example is shown in Fig. 6. This figure shows that the GA algorithm finds the global optimum in less than nearly 600 iterations. Note that this image has a bit more pixels 696 (29*24), the GA algorithm performs similar to other case studies. To see how the random images changed, the convergence curve is also given in Fig. 6. Gradual and steady convergence rate is

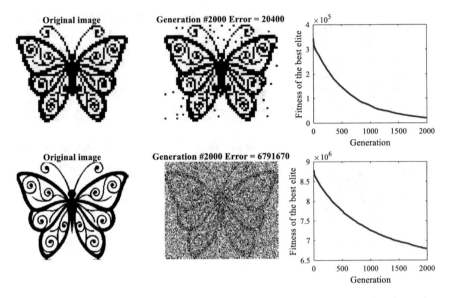

Fig. 8 The performance of GA with the same configuration degrades when increasing the number of variables (pixels) from 2784 (58*48) to 69600 (290*240)

evident again. To see how the initial random image is evolved, Fig. 7 shows the best elite found in iteration 1, 100, 200, 300, 400, 500, and 590 (where the algorithm stops). This figure shows significant improvement in the first 300 iterations where the algorithm explores and finds promising candidates in the exploitation phase. As the algorithm gets closer to the final iteration, the changes that lead to final tweaks in the image become smaller.

To see how much the performance of the GA algorithm degrades when changing the number of pixels (variable) of this problem, bigger images of the same butterfly with 2784 (58*48) and 69600 (290*240) pixels are used in the next experiment. Figure 8 shows the results of GA with the same configuration used in the preceding iterations. It can be seen that the performance of GA significantly degrades proportional to the number of variables. Increasing the number of chromosomes and/or iterations will influence the quality of solutions obtained by GA. Finding an optimal value for these parameters is outside the scope of this chapter. However, Fig. 9 is given to showcase the improvement in the case of increasing the number of chromosomes or generations.

Figure 9 shows that in case of improving either number of chromosomes or generations, the accuracy of the image constructed increases. However, it seems the accuracy is increased better when GA is given more generations that the number of chromosomes. This is not always the case on all problems since researchers normally count the number of function evaluations as a measure of cost. The discrepancy between both improvements is not significant, so we can say that tuning the number of chromosomes and iterations are both effective.

With 50 chromosomes and 2000 generations

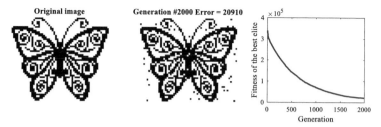

With 50 chromosomes and 4000 generations

With 100 chromosomes and 2000 generations

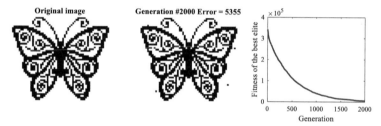

Fig. 9 Increasing the number of chromosomes of generations results in finding a more accurate image

Tuning other controlling parameters of GA can be helpful too. An experiment is conducted to see whether such modification is beneficial when solving the same problem. The probability of crossover, probability of mutation, and elitism ratio are changed in Fig. 10.

This figure shows that some of the changes is beneficial. Changing crossover to 1 requires all children to move to the next generation, which increases exploitation. The results in Fig. 10 that when setting the probability of crossover to its maximum value, better results can be obtained. This is due to the higher exploration that results in finding a more accurate estimation of the image.

Increasing the mutation rate leads to higher exploration since there are more random changes in the genes. The results in Fig. 10 show that increasing or decreasing this parameter should be done carefully since in the case of using very small or big values, the performance will degrade substantially.

50 chromosomes, 2000 generations, $P_c = 0.95$, $P_m = 0.001$, and ER = 0.2

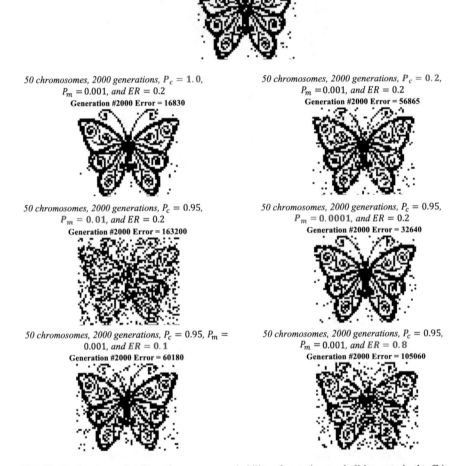

Fig. 10 Tuning the probability of crossover, probability of mutation, and elitism rate in the GA algorithm

Finally, Fig. 4 shows that the elitism ratio is also important in getting accurate results. This parameter indicates the portion of the best solution (elites) that should be kept and propagated to the next generation intact. Adding more elites results in more exploitation. By contrast, decreasing the number of elites causes higher exploration.

4   Conclusion

This chapter covered the main mechanisms of the genetic algorithm: initialization, selection, recombination, and mutation. The most widely used methods in each of these mechanism were discussed in details. Despite the focus on the original version of GA, the recent and new studies in each of the mechanisms were discussed too. The original version of GA is only able to solve problems with discrete variables, so the continuous version of this algorithm was discussed too. After the theoretical discussion and analysis, the GA algorithm was applied to a challenging problem in the field of image processing.

It was observed that the GA algorithm performed very well and constructed identical images when the number of pixels is not very large. The performance of GA degraded proportional to the number of pixels in the image. The chapter also investigated the impact of changing the controlling parameters on the performance of GA. It was observed that the number of chromosomes and generations play key roles. Other controlling parameters including probability of crossover, probability of mutation, and error ratio were investigated too. It was discussed that such parameters should be tuned very carefully (possibly with another optimizer) since they can significantly degrade the performance of GA in case of not using proper values.

References

1. Holland, J. H. (1992). Genetic algorithms. *Scientific American, 267*(1), 66–73.
2. Goldberg, D. E., & Holland, J. H. (1988). Genetic algorithms and machine learning. *Machine Learning, 3*(2), 95–99.
3. Genlin, J. (2004). Survey on genetic algorithm. *Computer Applications and Software, 2,* 69–73.
4. Cant-Paz, E. (1998). A survey of parallel genetic algorithms. *Calculateurs Paralleles, Reseaux et Systems Repartis, 10*(2), 141–171.
5. Goldberg, D. E., & Deb, K. (1991). A comparative analysis of selection schemes used in genetic algorithms. In *Foundations of Genetic Algorithms* (Vol. 1, pp. 69–93). Elsevier.
6. Goldberg, D. E. (1990). A note on Boltzmann tournament selection for genetic algorithms and population-oriented simulated annealing. *Complex Systems, 4*(4), 445–460.
7. Miller, B. L., & Goldberg, D. E. (1995). Genetic algorithms, tournament selection, and the effects of noise. *Complex Systems, 9*(3), 193–212.
8. Kumar, R. (2012). Blending roulette wheel selection & rank selection in genetic algorithms. *International Journal of Machine Learning and Computing, 2*(4), 365.
9. Syswerda, G. (1991). A study of reproduction in generational and steady-state genetic algorithms. In *Foundations of Genetic Algorithms* (Vol. 1, pp. 94–101). Elsevier.
10. Blickle, T., & Thiele, L. (1996). A comparison of selection schemes used in evolutionary algorithms. *Evolutionary Computation, 4*(4), 361–394.
11. Collins, R. J., & Jefferson, D. R. (1991). Selection in massively parallel genetic algorithms (pp. 249–256). University of California (Los Angeles). Computer Science Department.
12. Ishibuchi, H., & Yamamoto, T. (2004). Fuzzy rule selection by multi-objective genetic local search algorithms and rule evaluation measures in data mining. *Fuzzy Sets and Systems, 141*(1), 59–88.
13. Hutter, M. (2002, May). Fitness uniform selection to preserve genetic diversity. In *Proceedings of the 2002 Congress on Evolutionary Computation, CEC'02.* (Vol. 1, pp. 783–788). IEEE.

14. Grefenstette, J. J. (1989). How genetic algorithms work: A critical look at implicit parallelism. In *Proceedings 3rd International Joint Conference on Genetic Algorithms (ICGA89)*.
15. Syswerda, G. (1989). Uniform crossover in genetic algorithms. In *Proceedings of the Third International Conference on Genetic Algorithms* (pp. 2–9). Morgan Kaufmann Publishers.
16. Srinivas, M., & Patnaik, L. M. (1994). Genetic algorithms: A survey. *Computer, 27*(6), 17–26.
17. Semenkin, E., & Semenkina, M. (2012, June). Self-configuring genetic algorithm with modified uniform crossover operator. In *International Conference in Swarm Intelligence* (pp. 414–421). Springer, Berlin, Heidelberg.
18. Hu, X. B., & Di Paolo, E. (2007, September). An efficient genetic algorithm with uniform crossover for the multi-objective airport gate assignment problem. In *IEEE Congress on Evolutionary Computation CEC 2007* (pp. 55–62). IEEE.
19. Tsutsui, S., Yamamura, M., & Higuchi, T. (1999, July). Multi-parent recombination with simplex crossover in real coded genetic algorithms. In *Proceedings of the 1st Annual Conference on Genetic and Evolutionary Computation-Volume 1* (pp. 657–664). Morgan Kaufmann Publishers Inc.
20. Blickle, T., Fogel, D. B., & Michalewicz, Z. (Eds.). (2000). *Evolutionary computation 1: Basic algorithms and operators* (Vol. 1). CRC Press.
21. Oliver, I. M., Smith, D., & Holland, J. R. (1987). Study of permutation crossover operators on the travelling salesman problem. In *Genetic Algorithms and Their Applications: Proceedings of the Second International Conference on Genetic Algorithms* July 28–31. (1987). *at the Massachusetts institute of technology* (p. 1987) Cambridge, MA. Hillsdale, NJ: L. Erlhaum Associates.
22. Davis, L. (1985, August). Applying adaptive algorithms to epistatic domains. In *IJCAI* (Vol. 85, pp. 162–164).
23. Whitley, D., Timothy, S., & Daniel, S. Schedule optimization using genetic algorithms. In L Davis, (ed.), pp. 351–357.
24. Grefenstette, J., Gopal, R., Rosmaita, B., & Van Gucht, D. (1985, July). Genetic algorithms for the traveling salesman problem. In *Proceedings of the First International Conference on Genetic Algorithms and Their Applications* (pp. 160–168).
25. Louis, S. J., & Rawlins, G. J. (1991, July). Designer genetic algorithms: Genetic algorithms in structure design. In *ICGA* (pp. 53–60).
26. Eshelman, L. J., Caruana, R. A., & Schaffer, J. D. (1989, December). Biases in the crossover landscape. In *Proceedings of the Third International Conference on Genetic Algorithms* (pp. 10–19). Morgan Kaufmann Publishers Inc.
27. Deep, K., & Thakur, M. (2007). A new mutation operator for real coded genetic algorithms. *Applied Mathematics and Computation, 193*(1), 211–230.
28. Srinivas, M., & Patnaik, L. M. (1994). Adaptive probabilities of crossover and mutation in genetic algorithms. *IEEE Transactions on Systems, Man, and Cybernetics, 24*(4), 656–667.
29. Neubauer, A. (1997, April). A theoretical analysis of the non-uniform mutation operator for the modified genetic algorithm. In *IEEE International Conference on Evolutionary Computation, 1997* (pp. 93–96). IEEE.
30. Hinterding, R. (1995, November). Gaussian mutation and self-adaption for numeric genetic algorithms. In *IEEE International Conference on Evolutionary Computation, 1995* (Vol. 1, p. 384). IEEE.
31. Tsutsui, S., & Fujimoto, Y. (1993, June). Forking genetic algorithm with blocking and shrinking modes (fGA). In *ICGA* (pp. 206–215).
32. Oosthuizen, G. D. (1987). Supergran: a connectionist approach to learning, integrating genetic algorithms and graph induction. In *Genetic Algorithms and Their Applications: Proceedings of the Second International Conference on Genetic Algorithms: July 28–31. at the Massachusetts Institute of Technology* (p. 1987) Cambridge, MA. Hillsdale, NJ: L. Erlhaum Associates.
33. Mauldin, M. L. (1984, August). Maintaining diversity in genetic search. In *AAAI* (pp. 247–250).
34. Ankenbrandt, C. A. (1991). An extension to the theory of convergence and a proof of the time complexity of genetic algorithms. In *Foundations of genetic algorithms* (Vol. 1, pp. 53–68). Elsevier.

35. Ahn, C. W., & Ramakrishna, R. S. (2003). Elitism-based compact genetic algorithms. *IEEE Transactions on Evolutionary Computation, 7*(4), 367–385.
36. Zitova, B., & Flusser, J. (2003). Image registration methods: A survey. *Image and Vision Computing, 21*(11), 977–1000.

Grey Wolf Optimizer: Theory, Literature Review, and Application in Computational Fluid Dynamics Problems

Seyedali Mirjalili, Ibrahim Aljarah, Majdi Mafarja,
Ali Asghar Heidari and Hossam Faris

Abstract This chapter first discusses inspirations, methematicam models, and an in-depth literature of the recently proposed Grey Wolf Optimizer (GWO). Then, several experiments are conducted to analyze and benchmark the performance of different variants and improvements of this algorithm. The chapter also investigates the application of the GWO variants in finding an optimal design for a ship propeller.

1 Introduction

Over the course of last decade, stochastic optimization techniques superseded conventional, deterministic approaches due to several reasons. One of the main reasons is the inefficiency of deterministic optimization or search methods for solving NP hard problems. In fact, for a lot of NP-hard problems, there is no deterministic solution

S. Mirjalili (✉)
Institute of Integrated and Intelligent Systems, Griffith University, Nathan,
Brisbane, QLD 4111, Australia
e-mail: seyedali.mirjalili@griffithuni.edu.au

I. Aljarah · H. Faris
King Abdullah II School for Information Technology, The University of Jordan,
Amman, Jordan
e-mail: i.aljarah@ju.edu.jo

H. Faris
e-mail: hossam.faris@ju.edu.jo

M. Mafarja
Department of Computer Science, Faculty of Engineering and Technology,
Birzeit University, PoBox 14, Birzeit, Palestine
e-mail: mmafarja@birzeit.edu

A. A. Heidari
School of Surveying and Geospatial Engineering, University of Tehran, Tehran, Iran
e-mail: as_heidari@ut.ac.ir

© Springer Nature Switzerland AG 2020
S. Mirjalili et al. (eds.), *Nature-Inspired Optimizers*, Studies in Computational
Intelligence 811, https://doi.org/10.1007/978-3-030-12127-3_6

87

as of toady. Another reason is the dependency of some of deterministic algorithms on calculating the derivation of the problem. For the problems that derivation is ill-defined or difficult to obtain, such methods cannot be used.

On the other hand, stochastic optimization techniques are able to find near optimal solutions for NP-hard problems in a reasonable time. Also, the majority of them consider optimization problems as a black box and do not require derivation. This means that the same algorithm can be applied to different problems without the need to know the internal mathematical model or the computer program of an optimization problem.

A set of popular stochastic optimization algorithms that have been very popular lately include nature-inspired techniques. Such methods mimic natural intelligence and provide nature-inspired problem solving techniques. One of the seminal algorithms in this area is the well-regarded Genetic Algorithm (GA). This algorithm has been inspired from the biological evolution and mimics the process of evolving creates in a computer. In fact, this algorithm has been equipped with selection, recombination, and mutation operators to do this. GA belongs to the family of evolutionary algorithms. Other popular evolutionary techniques are Evolution Strategy (EA) and Differential Evolution (DE).

After the proposal of GA, several other classes of nature-inspired algorithms came to the existence. One of the classes included Swam Intelligence, in which the social and collective intelligence of organisms in nature were the main inspiration. For instance, Ant Colony Optimization (ACO) mimics the intelligent method of ants in finding the nearest path from a nest to a food source. Other popular algorithms in this class are Particle Swarm Optimization (PSO) and Artificial Bee Colony (ABC) algorithm. Swarm intelligence algorithms have been used in several applications.

One of the recent and well-regarded swarm intelligence techniques is the Grey Wolf Optimizer (GWO). This algorithm simulated the social hierarchy of intelligence of grey wolves in hunting. This algorithm was proposed by Mirjalili et al. in 2014 [1] and has been widely used in both science and industry ever since. This chapter presents the preliminaries and essential definitions of this algorithm and analyzes its performance. The rest of the chapter is organized as follows:

The inspiration and mathematical model of GWO are presented in Sect. 2. Section 3 provides an brief literature review of GWO. The performance of GWO is analyzed on a set of test beds in Sect. 4. Section 5 applies GWO to CFD problems. Finally, Sect. 6 concludes the chapter.

2 Grey Wolf Optimizer

The GWO algorithm mimics the hierarchy and social interaction of grey wolves in nature. In a pack of wolves, there are different types of members considering the domination level including α, β, δ, and ω. The most dominant wolf is α and the domination and leadership power decrease from α to ω (Fig. 1).

Fig. 1 Domination hierarchy of grey wolves in a pack

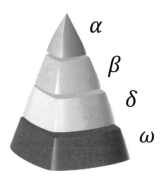

This mechanism is implemented by dividing a population of candidate solutions for a given optimization problem into four classes. This process is shown in Fig. 2 for a population of six solutions. As can be seen in Fig. 2, the first three best solutions are considered as α, β, and δ. The rest of solutions are included in the group of ω wolves. To implement this, we need to update the hierarchy in every iteration before changing the solutions. After this division, the position of solutions are updated using the following mathematical models:

2.1 Encircling Prey

Wolves normally hunt in a pack. This means that they collaborate in an intelligent manner to catch and consume a prey. Grey wolves first chase the prey in a team and try to encircle it, change its movement direction, and increase the chance of hunt. An example of the process of simulating this behavior in a computer is shown in Fig. 4 in one dimension (Fig. 3).

Fig. 2 Devision of the population into four classes in GWO

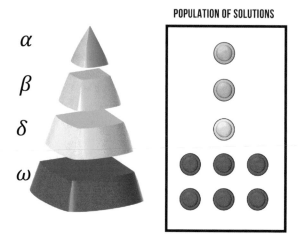

POPULATION OF SOLUTIONS

Fig. 3 The position of a wolf can be defined around a prey with considering a 1D mesh around the prey

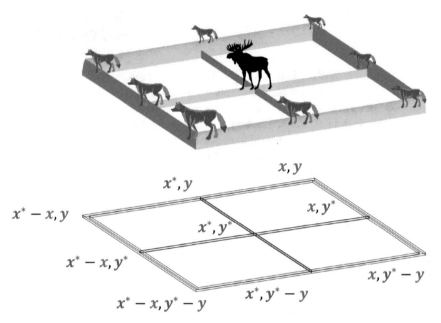

Fig. 4 The position of a wolf can be defined around a prey with considering a 2D mesh around the prey

The same concept can be applied when considering y or z axes.

This figure shows that different positions can be achieved when considering the location of the grey wolf (w, y) and the location of the prey (x^*, y^*). This is can be simulated for any number of dimensions, and Fig. 5 is an example.

This has been mathematically modeled using the following equation.

$$\vec{X}(t+1) = \vec{X_p}(t) - \vec{A} \cdot \vec{D} \tag{1}$$

where $\vec{X}(t+1)$ is the next location of the wolf, $\vec{X}(t)$ is current location, $\vec{A}$ is a coefficient and $\vec{D}$ is the distance that depends on the location of the prey $(\vec{X_p})$ and is calculated as follows:

$$\vec{D} = \left| \vec{C} \cdot \vec{X_p}(t) - \vec{X}(t) \right| \tag{2}$$

Fig. 5 Different positions of
a grey wolf around a prey

If we merge these two equations, the following equation can be written:

$$\vec{X}(t+1) = \vec{X}_p(t) - \vec{A} \cdot \left| \vec{C} \cdot \vec{X}_p(t) - \vec{X}(t) \right| \tag{3}$$

With these two equations (or the combined one), a solution is able to relocate around another solution. Since X is a vector, we can add any number of dimensions. The random components in the above equations are essential since without them the grey wolf moves to a fixed number of positions around the prey (1 in 1D space, 7 in a 2D space, and 25 in a 3D space).

The equations for defining the random components are as follows:

$$\vec{A} = 2\vec{a} \cdot \vec{r}_1 - \vec{a} \tag{4}$$

$$\vec{C} = 2 \cdot \vec{r}_2 \tag{5}$$

where $\vec{a}$ is a parameter that decreases linearly from 2 to 0 during optimization. $\vec{r}_1$ and $\vec{r}_2$ are randomly generated from the interval [0,1]. The equation to update the parameter a is as follows:

$$a = 2 - t \left(\frac{2}{T} \right) \tag{6}$$

where t shows the current iteration and T is the maximum number of iterations.

2.2 Hunting the Prey

The above equations allow a wolf to relocate to any points in a hyper-shpere around the prey. However, this is not enough to simulate to social intelligence of grey wolves. To simulate the prey, we assume that the best solution found so far (which is the alpha wolf) is the position of the prey. This is an essential assumption because in real-world problems, we do not know the position of the global optimum. Therefore, we assume that the alpha knows the position of the prey since he is leading the pack.

With defining hierarchy, developing an equation for encircle, and defining the position of prey, the position of each wolf can be updated using the following equations:

$$\vec{X}(t+1) = \frac{X_1 + X_2 + X_3}{3} \tag{7}$$

where $\vec{X_1}$ and $\vec{X_2}$ and $\vec{X_3}$ are calculated with Eq. 8.

$$\begin{aligned}
\vec{X}_1 &= \vec{X}_\alpha(t) - \vec{A_1} \cdot \vec{D_\alpha} \\
\vec{X}_2 &= \vec{X}_\beta(t) - \vec{A_2} \cdot \vec{D_\beta} \\
\vec{X}_3 &= \vec{X}_\delta(t) - \vec{A_3} \cdot \vec{D_\delta}
\end{aligned} \tag{8}$$

$\vec{D_\alpha}$, $\vec{D_\beta}$ and $\vec{D_\delta}$ are calculated using Eq. 9.

$$\begin{aligned}
\vec{D_\alpha} &= \left| \vec{C_1} \cdot \vec{X_\alpha} - \vec{X} \right| \\
\vec{D_\beta} &= \left| \vec{C_2} \cdot \vec{X_\beta} - \vec{X} \right| \\
\vec{D_\delta} &= \left| \vec{C_3} \cdot \vec{X_\delta} - \vec{X} \right|
\end{aligned} \tag{9}$$

3 Literature Review

The GWO algorithm was proposed in 2014 [1]. As of mid 2018, this algorithm has been cited 1100 times according to Google Scholar metrics. The works after GWO can be divided into three main classes: different variants, improved versions, and applications. This section covers some of the most well-regarded works in each of these classes.

3.1 Different Variants of GWO

The original version of GWO has been designed to solve problems with no constraints and only one objective, so it can be considered as an unconstrained single-objective optimization algorithm. Without modification, this algorithm can be applied to a wide range of problems. To solve problems of different types, however, the GWO algorithm requires some modifications.

3.1.1 Discrete GWO

The original version of GWO is able to solve problems with continuous variables, so it requires modifications to solve binary or discrete problems. There are several works in the literature as follows:

In 2016, Emary et al. [2] proposed the use of evolutionary operators (crossover) to solve binary problems using GWO. A similar evolutionary operators are utilized in [3] too. To solve economic dispatch problems, Jayabarathi used both crossover and mutation operators to combine the artificial grey wolves instead of the normal position updating equation of this algorithm. [4]. In 2017, a quantum-inspired binary version of the GWO algorithm was proposed as well [5]. In this technique, the hierarchy levels are replaced by Q-bit's individual probabilistic representation. The authors also used dynamic rotation angle and coordinate rotation gate as well. A similar work was conducted by Sujatha and Punithavathani [6], in which a Genetic Grey Wolf Optimizer (GGWO) was proposed. A multi-objective discrete GWO was proposed by Lu et al. in 2016 [7], in which an encoding strategy was used with modification in the mathematical model of searching for the prey in GWO. Transfer functions are used in the GWO algorithm by Wang et al. in [8]. Other binary variants can be found in [9–13]

3.1.2 Constrained GWO

As discussed above, there is no constraint handling technique in the original GWO. Some of the popular methods to solve constrained problems using GWO are discussed in this subsection.

Chaotic GWO was integrated with a simple penalty function and applied to a number of constrained engineering design problems [14]. To solve optimal power problems with static and dynamic constraints, Teeparthi and Kumar [15] used several penalty functions to handle both inequality and equality constraints. Gupta and Deep integrated a random walk into GWO and employed a barrier penalty function to solve constrained problems with the GWO algorithm [16]. Penalty functions are also used in [18–20, 22] Another different approach for handling constraints using GWO was proposed by Yang and Long [17], in which a set of feasibility-based rules based on tournament selection was introduced to handle constraints [21].

3.1.3   Multi-objective GWO

The first multi-objective version of GWO was proposed by Mirjalili et al. in 2015 [23]. The main operators and mechanisms employed were very similar to those of Multi-Objective Particle Swarm Optimization (MOPSO) [24]: a archive to store non-dominated Pareto optimal solutions and an archive handler to manage the solutions in the archive.

In 2017, a hybrid multi-objective grey wolf optimizer (HMOGWO) was proposed for dynamic scheduling in a real-world welding industry [25]. In the HMOGWO, the social hierarchy of wolves in MOGWO was modified to further improve its exploration and exploitation. Another improvement was the use of genetic operators in MOGWO. In 2018, another hybrid multi-objective GWO was proposed [26]. This method uses a dynamic maintenance strategy for the archive in GWO and a modified leader selection mechanism. There were also other mortifications and integrations including generic initial population with high-quality solutions and a local search. In 2018, a non-dominated sorting multi-objective Grey Wolf Optimizer was proposed [27]. Another MOGWO based on non-dominated sorting can be found in [28]. The binary version of MOGWO has been proposed in the literature too in [29]. In this method, an encoding/decoding method was proposed. For the social hierarchy mechanism, a non-dominated sorting operator is used.

3.1.4   Hybrid GWO

In the literature of heuristics and meta-heuristics, hybridization of multiple algorithms is one way to benefit from the advantages of different algorithms. There has been a large number of hybrid algorithms using GWO as one of the algorithms in the process of hybridization. Some of such hybrids are as follows:

- Hybrid GWO and PSO [30–34]
- Hybrid GWO and GA [35]
- Hybrid GWO and ACO [36]
- Hybrid GWO and BAT [37, 38]
- Hybrid GWO and Flower Pollination Algorithm (FPA) [39]
- Hybrid GWO and DE [40]
- Hybrid GWO and SCA [41]
- Hybrid GWO and BBO [42]

4   Performance Analysis of GWO

In this section, the performance of the GWO algorithm is tested on several test functions. In the field of optimization, test functions are usually used to analyze the performance of optimization algorithms due to two main reasons. Firstly, the global optimum is known in a test function, hence, we can quantitatively calculate the discrepancy between a solution obtained by an algorithm and the global optimum of the test functions. This is not the case for real-world problems because the global optimum of such problems are normally unknown.

Secondly, test functions have mostly known shapes for their search landscapes and there are controlling parameters to increase their difficulties. Therefore, the performance of an algorithm can be benchmarked from different perspectives and diverse levels of difficulties.

A set of test functions including unimodal and multi-modal is chosen here. The main controlling parameters in these benchmark functions is the number of variables. The GWO algorithm solves the 30-dimensional versions and the results are presented and discussed in the following paragraphs.

4.1 Convergence Behaviour of GWO

One of the most popular methods to analyze the performance of stochastic algorithms is to visualize the best solution found so far over the course of generations, which is called the convergence curve. The convergence curve of GWO when solving two unimodal and two multi-modal test functions are visualized in Fig. 6. Note that 500 iterations and 30 solutions are used in this experiment. Also, each test function has 30 variables to be optimized by GWO.

This figure shows that the GWO algorithm shows a fast convergence rate on the test functions. In the first test function, the convergence rate is constant over the course of iterations. Considering the shape of this test function, one reason is that the GWO finds a better alpha, beta, and delta in each iteration due to the unimodal nature of the problem since the slop towards the global optimum does not change. This drives other solutions towards the global optimum.

In the second unimodal test function, there is a sharp decline at the beginning of the optimization. This can be explained by looking at the shape of the test function in Fig. 6. The slope is very steep towards the global optimum, so GWO is likely to find much better solutions as compared to the first initial population. After this stage, the convergence curve is steady since the basin of the test function is almost flat.

The convergence curve of the GWO on multi-modal test functions is slightly different, in which there is a small improvement in the solutions in the final stages of optimization. Due to a large number of local solutions in such search spaces, the search is more difficult and the solutions tend to be trapped in local optima. This requires more computational resources to avoid them and eventually find the global optimum. As a result, the convergence rate tends to be slower on multi-modal test functions than unimodal ones.

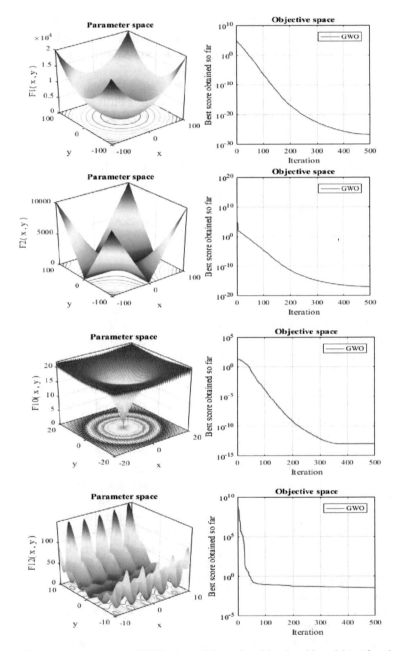

Fig. 6 Convergence behaviour of GWO when solving unimodal and multi-modal test functions

4.2   Changing the Main Controlling Parameter of GWO

The main controlling parameter in the GWO is $\vec{A}$, which is calculated as follows:

$$\vec{A} = 2\vec{a} \cdot \vec{r}_1 - \vec{a} \tag{10}$$

$$\vec{a} = 2 - \vec{t}\left(\frac{2}{T}\right) \tag{11}$$

where t shows the current iteration and T is the maximum number of iterations.

With tuning this parameter, different exploratory and exploitative search patterns can be achieved. In this subsection, the constant used in this equation is first parametrized as follows:

$$\vec{a} = a_{ini} - \vec{t}\left(\frac{a_{ini}}{T}\right) \tag{12}$$

where t shows the current iteration and T is the maximum number of iterations.

Eight different values including 0.5, 1, 1.5, 2, 2.5, 3, 3.5, and 4 are used to see how the performance of GWO can be changed based on these values. As an example, the results of GWO when solving the first test function are visualized in Fig. 7.

This figure shows very interesting patterns. In the original version of the GWO algorithm, a_{ini} was set to 2. In the original paper fo GWO, it is claimed that it provides a good balance between exploration and exploitation. In Fig. 7, it can be seen that when a_{ini} is equal to 0.5, 1, and 1.5, the convergence curves of the GWO algorithm show faster convergence rates. By contrast, the convergence rate becomes slower as a_{ini} is increased.

Fig. 7 Convergence behaviour of GWO when changing a_{ini} in the vector A

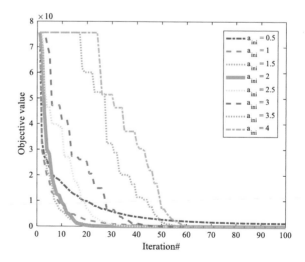

The impact of these behaviours on the exploration and exploitation of GWO can be seen in Fig. 8. This figure shows that the distance between solutions is less when $a_{ini} < 2$. In fact, the average distance between solutions never increases, which shows pure exploitation of the search space using the algorithm. As a_{ini} increases, however, the exploration is more since the diversity becomes higher. This indicates the exploration of the search space.

The test function used in this subsection was F1, which is unimodal. To see whether the algorithm shows the same behavior on multimodal test functions or not, Fig. 9 is given. This figures show similar behaviours in both convergence rate and diversity.

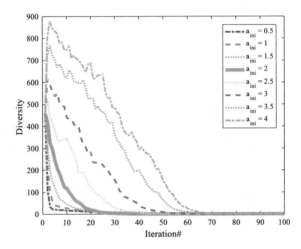

Fig. 8 Average distances (diversity) between solutions in GWO when changing a_{ini} in the vector A. It is evident that exploration is increased proportional to the value of a_{ini}

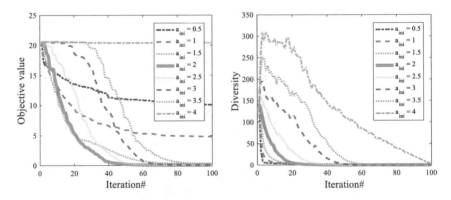

Fig. 9 Average distances (diversity) between solutions in GWO when changing a_{ini} in the vector A on a multimodal test function. It is evident that exploration is increased proportional to the value of a_{ini}

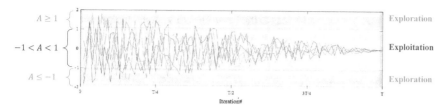

Fig. 10 Behaviour of parameter A and its impact on exploration and exploitation of the GWO

The behavior of the parameter A is also visualized in Fig. 10. This figure shows that this parameter is stochastic, which is due to the random component in its equation. Although its value changes randomly, the parameter a changes its range so that it shows different exploratory and exploitative behaviours. The GWO shows exploration when $A \geq 1$ or $A \leq 1$. By contrast, the GWO algorithm exploit the search space when $-1 < A < 1$.

5 Marine Propeller Design Using GWO

In this section, a challenging real-world problem (ship propeller design) is tackled by the GWO algorithm. Since this problem is constrained, a constraint handling mechanism is required. Finding the best constraint handling method is outside the scope of this chapter. Therefore, a barrier function is applied that penalizes the solutions that violate the constraints at any level with a large objective value since both case studies are minimization problems.

In this problem, an optimal design for a ship propeller with five blades and 2 meters diameter should be found. The shape of this propeller designed by a designer using trial and error is shown in Fig. 11.

There are two objectives for this problem: maximizing efficiency and minimizing cavitation. The 'efficiency' refers to the effectiveness of a propeller to cover engines'

Fig. 11 The ship propeller design case study considered in this chapter

2 meters

power to thrust. Cavitation refers to the formation of vapour cavities in a liquid (water in this case study) when the pressure changes significantly during the operation a propeller. Such cavitates create micro shock waves that damage the propeller in long-term. Since the main focus of this chapter and book is on single-objective optimization, both objectives are optimized independently. Therefore, the problem formulation can be as follows:

$$\text{Maximize: } \eta(\overrightarrow{X}) \tag{13}$$

$$\text{Minimize: } Cav(\overrightarrow{X}) \tag{14}$$

$$\text{Subject to: } T > 40000, RPM = 200, Z = 7, D = 2, d = 0.4, S = 5 \tag{15}$$

where η shows the efficiency, Cav indicates cavitation, $\overrightarrow{X}$ includes the structural parameters of the airfoils as discussed before, T is the desired thrust, RPM indicates the Revolution Per Minute, Z is the number of blades, D shows the number of blades in meters, d is the hub's diameter, and S shows the speed of ship (m/s).

In the above equation, the vector $\overrightarrow{X}$ includes all the parameters for this problem. The main structural parameters of the propeller design problem considered in this chapter are shown in Fig. 12. It can be seen that the shape of each blade can be defined using multiple airfoils. The blades are also identical, so one blade should be optimized. In order to find the final geometrical shape of the blade, standard rules of National Advisory Committee for Aeronautics (NACA) is used.

In this chapter, 10 airfoils are considered and as discussed above, each airfoil requires two parameters, hence, there are total number of 20 parameters to be optimized. Note that the full list of constraints and other details about this problem can be found in [45].

As discussed above, the objectives are considered separately due to the scope of this chapter and book. Therefore, two experiments are conducted. In the first experiment, the efficiency is maximized for three propellers with five, six, and seven blades. In the second experiment, the cavitation is minimized for three propellers with five,

Fig. 12 The structural parameters of the propeller design problem

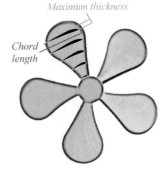

six, and seven blades. The GWO algorithm tackles this problem using 30 number of solutions and 500 number of iterations. The results are shown in Figs. 13 and 14.

It can be seen in these results that the GWO algorithm does not converge in more than half of the iterations when maximizing the efficiency and then, suddenly starts converging. This is due to the large number of constraints in this problem, in which all solutions are infeasible for many iterations. When all solutions are infeasible, the main operators of GWO are not effective and the algorithm keeps generating random numbers for finding feasible solutions to start the search process in the feasible regions of the search space. This can be improved by giving the algorithm some feasible solutions to start with or tunning the controlling parameters of GWO. However, such experiments are outside the scope of this chapter. It is also interesting that the best efficiency is found for a five-blade propeller.

Similar patterns can be observed in Fig. 14 when minimizing the cavitation. The algorithm starts with infeasible solutions until nearly two-thirds of the iterations. After this point, it seems that the algorithm finds a feasible solution. The curves show that convergence rates are very high and the algorithm accelerates towards

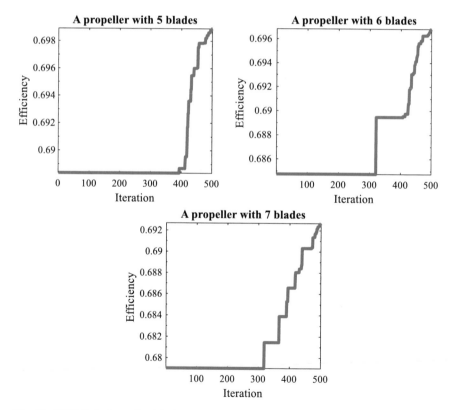

Fig. 13 GWO finds the optimal design parameters for propellers with different number of blades when maximizing the efficiency measure

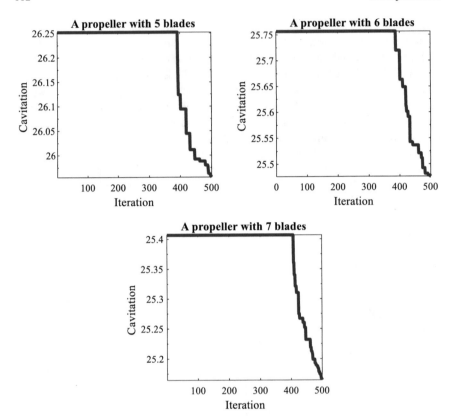

Fig. 14 GWO finds optimal designs for propellers with different number of blades when minimizing the cavitation

the global optimum. Increasing the number of solutions and iterations can improve these results. However, the solutions found by the GWO algorithm already show high efficiency and low cavitation. Finding the best possible solution requires running GWO multiple times and fine tuning, which is beyond the scope of this chapter. It is also interesting that the best deficiency is found for a seven-bladed propeller. This is in contrast to what was observed when maximizing the efficiency measure. Therefore, these results show that if a designer is more concern about efficiency, fewer number of blades should be made. By contrast, if the aviation is more important, more blades should be designed and made.

6 Conclusions

This chapter provided an in-depth literature review and analysis of the GWO algorithm as one of the recent population-based optimizers. Different variants of this algorithm were covered including binary, multi-objective, constrained, and hybridized.

GWO was applied to several test functions for observing its performance. The exploratory and exploitative search patterns of this algorithm were observed using several diagrams including diversity and convergence curves. The chapter also investigated the application of GWO in finding an optimal design for a ship propeller.

The GWO algorithm has been widely used since its proposal and proven to be very effective for challenging real-world problems. This chapter confirms the merits of this algorithm once again.

References

1. Mirjalili, S., Mirjalili, S. M., & Lewis, A. (2014). Grey wolf optimizer. *Advances in Engineering Software, 69,* 46–61.
2. E., Zawbaa, H. M., & Hassanien, A. E. (2016). Binary grey wolf optimization approaches for feature selection. *Neurocomputing, 172,* 371–381.
3. Panwar, L. K., Reddy, S., Verma, A., Panigrahi, B. K., & Kumar, R. (2018). Binary Grey Wolf Optimizer for large scale unit commitment problem. *Swarm and Evolutionary Computation, 38,* 251–266.
4. Jayabarathi, T., Raghunathan, T., Adarsh, B. R., & Suganthan, P. N. (2016). Economic dispatch using hybrid grey wolf optimizer. *Energy, 111,* 630–641.
5. Srikanth, K., Panwar, L. K., Panigrahi, B. K., Herrera-Viedma, E., Sangaiah, A. K., & Wang, G. G. (2017). Meta-heuristic framework: Quantum inspired binary grey wolf optimizer for unit commitment problem. *Computers & Electrical Engineering.*
6. Sujatha, K., & Punithavathani, D. S. (2018). Optimized ensemble decision-based multi-focus imagefusion using binary genetic Grey-Wolf optimizer in camera sensor networks. *Multimedia Tools and Applications, 77*(2), 1735–1759.
7. C., Xiao, S., Li, X., & Gao, L. (2016). An effective multi-objective discrete grey wolf optimizer for a real-world scheduling problem in welding production. *Advances in Engineering Software, 99,* 161–176.
8. Wang, S., Hua, G., Hao, G., & Xie, C. (2017). A comparison of different transfer functions for binary version of grey wolf optimiser. *International Journal of Wireless and Mobile Computing, 13*(4), 261–269.
9. L., Sun, L., Guo, J., Qi, J., Xu, B., & Li, S. (2017). Modified discrete grey wolf optimizer algorithm for multilevel image thresholding. *Computational intelligence and neuroscience.*
10. Seth, J. K., & Chandra, S. (2016, March). Intrusion detection based on key feature selection using binary GWO. In *2016 3rd International Conference on Computing for Sustainable Global Development (INDIACom)* (pp. 3735–3740). IEEE.
11. Manikandan, S. P., Manimegalai, R., & Hariharan, M. (2016). Gene Selection from microarray data using binary grey wolf algorithm for classifying acute leukemia. *Current Signal Transduction Therapy, 11*(2), 76–83.
12. Li, L., Sun, L., Kang, W., Guo, J., Han, C., & Li, S. (2016). Fuzzy multilevel image thresholding based on modified discrete grey wolf optimizer and local information aggregation. *IEEE Access, 4,* 6438–6450.
13. Reddy, S., Panwar, L. K., Panigrahi, B. K., & Kumar, R. (2016, December). Optimal scheduling of uncertain wind energy and demand response in unit commitment using binary grey wolf optimizer (BGWO). In *2016 IEEE Uttar Pradesh Section International Conference on Electrical, Computer and Electronics Engineering (UPCON)* (pp. 344–349). IEEE.
14. Kohli, M., & Arora, S. (2017). Chaotic grey wolf optimization algorithm for constrained optimization problems. *Journal of Computational Design and Engineering.*
15. Teeparthi, K., & Kumar, D. V. (2016, December). Grey wolf optimization algorithm based dynamic security constrained optimal power flow. In *Power Systems Conference (NPSC), 2016 National* (pp. 1–6). IEEE.

16. Gupta, S., & Deep, K. Random walk grey wolf optimizer for constrained engineering optimization problems. *Computational Intelligence*.

17. Yang, J. C., & Long, W. (2016). Improved grey wolf optimization algorithm for constrained mechanical design problems. *Applied Mechanics and Materials, 851*, 553–558). Trans Tech Publications.

18. Joshi, H., & Arora, S. (2017). Enhanced grey wolf optimisation algorithm for constrained optimisation problems. *International Journal of Swarm Intelligence, 3*(2–3), 126–151.

19. Prakasam, S., Venkatachalam, M., & Saroja, M. (2016). Grey Wolf optimizer for constrained hardware-software codesign partitioning. *Programmable Device Circuits and Systems, 8*(8), 239–243.

20. Kumar, G., & Ranga, V. (2017, August). Meta-heuristic solution for relay nodes placement in constrained environment. In *2017 Tenth International Conference on Contemporary Computing (IC3)* (pp. 1–6). IEEE.

21. Long, W., Liang, X., Cai, S., Jiao, J., & Zhang, W. (2017). A modified augmented Lagrangian with improved grey wolf optimization to constrained optimization problems. *Neural Computing and Applications, 28*(1), 421–438.

22. Sreenu, K., & Malempati, S. (2017). MFGMTS: Epsilon constraint-based modified fractional grey wolf optimizer for multi-objective task scheduling in cloud computing. *IETE Journal of Research*, 1–15.

23. Mirjalili, S., Saremi, S., Mirjalili, S. M., & Coelho, L. D. S. (2016). Multi-objective grey wolf optimizer: a novel algorithm for multi-criterion optimization. *Expert Systems with Applications, 47*, 106–119.

24. Coello, C. A. C., Pulido, G. T., & Lechuga, M. S. (2004). Handling multiple objectives with particle swarm optimization. *IEEE Transactions on Evolutionary Computation, 8*(3), 256–279.

25. Lu, C., Gao, L., Li, X., & Xiao, S. (2017). A hybrid multi-objective grey wolf optimizer for dynamic scheduling in a real-world welding industry. *Engineering Applications of Artificial Intelligence, 57*, 61–79.

26. Yang, Z., & Liu, C. (2018). A hybrid multi-objective gray wolf optimization algorithm for a fuzzy blocking flow shop scheduling problem. *Advances in Mechanical Engineering, 10*(3), 1687814018765535.

27. Jangir, P., & Jangir, N. (2018). A new Non-Dominated Sorting Grey Wolf Optimizer (NS-GWO) algorithm: Development and application to solve engineering designs and economic constrained emission dispatch problem with integration of wind power. *Engineering Applications of Artificial Intelligence, 72*, 449–467.

28. Sahoo, A., & Chandra, S. (2017). Multi-objective Grey Wolf Optimizer for improved cervix lesion classification. *Applied Soft Computing, 52*, 64–80.

29. Lu, C., Xiao, S., Li, X., & Gao, L. (2016). An effective multi-objective discrete grey wolf optimizer for a real-world scheduling problem in welding production. *Advances in Engineering Software, 99*, 161–176.

30. Kamboj, V. K. (2016). A novel hybrid PSOGWO approach for unit commitment problem. *Neural Computing and Applications, 27*(6), 1643–1655.

31. Singh, N., & Singh, S. B. (2017). Hybrid algorithm of particle swarm optimization and Grey Wolf optimizer for improving convergence performance. *Journal of Applied Mathematics*.

32. Chopra, N., Kumar, G., & Mehta, S. (2016). Hybrid GWO-PSO algorithm for solving convex economic load dispatch problem. *International Journal Research Advanced Technology, 4*(6), 37–41.

33. Eid, H. F., & Abraham, A. (2018). Plant species identification using leaf biometrics and swarm optimization: A hybrid PSO, GWO, SVM model. *International Journal of Hybrid Intelligent Systems, (Preprint)*, 1–11.

34. Jain, U., Tiwari, R., & Godfrey, W. W. (2018). Odor source localization by concatenating particle swarm optimization and Grey Wolf optimizer. In *Advanced Computational and Communication Paradigms* (pp. 145–153). Springer, Singapore.

35. Tawhid, M. A., & Ali, A. F. (2017). A Hybrid grey wolf optimizer and genetic algorithm for minimizing potential energy function. *Memetic Computing, 9*(4), 347–359.

36. Ab Rashid, M. F. F. (2017). A hybrid Ant-Wolf Algorithm to optimize assembly sequence planning problem. *Assembly Automation, 37*(2), 238–248.
37. Abdelazeem, M. (2018, January). A hybrid Grey Wolf-bat algorithm for global optimization. In *The International Conference on Advanced Machine Learning Technologies and Applications (AMLTA2018)* (Vol. 723, p. 3). Springer.
38. ElGayyar, M., Emary, E., Sweilam, N. H., & Abdelazeem, M. (2018, February). A hybrid Grey Wolf-bat algorithm for global optimization. In *International Conference on Advanced Machine Learning Technologies and Applications* (pp. 3–12). Springer, Cham.
39. Pan, J. S., Dao, T. K., & Chu, S. C. (2017, November). A novel hybrid GWO-FPA algorithm for optimization applications. In *International Conference on Smart Vehicular Technology, Transportation, Communication and Applications* (pp. 274–281). Springer, Cham.
40. Debnath, M. K., Mallick, R. K., & Sahu, B. K. (2017). Application of hybrid differential evolution Grey Wolf optimization algorithm for automatic generation control of a multi-source interconnected power system using optimal fuzzy PID controller. *Electric Power Components and Systems, 45*(19), 2104–2117.
41. Singh, N., & Singh, S. B. (2017). A novel hybrid GWO-SCA approach for optimization problems. *Engineering Science and Technology, an International Journal.*
42. Zhang, X., Kang, Q., Cheng, J., & Wang, X. (2018). A novel hybrid algorithm based on Biogeography-based optimization and Grey Wolf optimizer. *Applied Soft Computing, 67*, 197–214.
43. Mirjalili, S. (2016). SCA: A sine cosine algorithm for solving optimization problems. *Knowledge-Based Systems, 96*, 120–133.
44. Drela, M. (1989). XFOIL: An analysis and design system for low Reynolds number airfoils. In *Low Reynolds number aerodynamics* (pp. 1–12). Springer, Berlin, Heidelberg.
45. Carlton, J. (2012). Marine propellers and propulsion. Butterworth-Heinemann.

Grasshopper Optimization Algorithm: Theory, Literature Review, and Application in Hand Posture Estimation

Shahrzad Saremi, Seyedehzahra Mirjalili, Seyedali Mirjalili
and Jin Song Dong

Abstract This chapter covers the fundamental concepts of the recently proposed Grasshopper Optimization Algorithm (GOA). The inspiration, mathematical model, and the algorithm are presented in details. A brief literature review of this algorithm including different variants, improvement, hybrids, and applications are given too. The performance of GOA is tested on a set of test functions including unimodal, multi-modal, and composite. The results show the ability of GOA in improving the quality of a random population, transiting from exploration to exploitation, showing high coverage of the search space, and accelerating the convergence curve over the course of iterations. The chapter also applies the GOA algorithm to a challenging problem in the field of hand posture estimation. It is observed that GOA finds an accurate configuration for a 3D hand model to match a given hand image acquired from a camera.

S. Saremi · S. Mirjalili (✉) · J. S. Dong
Institute of Integrated and Intelligent Systems, Griffith University,
Nathan, Brisbane, QLD 4111, Australia
e-mail: seyedali.mirjalili@griffithuni.edu.au

S. Saremi
e-mail: shahrzad.saremi@griffithuni.edu.au

J. S. Dong
Department of Computer Science, School of Computing, National University of Singapore,
Singapore, Singapore
e-mail: j.dong@griffith.edu.au

S. Mirjalili
School of Electrical Engineering and Computing, University of Newcastle,
Callaghan, NSW 2308, Australia
e-mail: sz.mirjalili@gmail.com

© Springer Nature Switzerland AG 2020
S. Mirjalili et al. (eds.), *Nature-Inspired Optimizers*, Studies in Computational
Intelligence 811, https://doi.org/10.1007/978-3-030-12127-3_7

1 Introduction

Swarm Intelligence (SI) [1] is a field that deals with understanding and simulating the collective behaviours of organisms without a centralized control unit in nature. In this field, it is assumed that a population is made of simple agents that interact with each other and/or environment. Such agents start to interact locally and often incorporate random mechanisms to achieve a goal globally. Some of the examples in nature are: ant colony, school of fish, a flock of birds, and a herd of buffaloes.

As an example, Raynold [2] simulated the swarming behaviour of birds using three simple operators: alignment, separation, and cohesion as can be seen in Fig. 1. Given the fact that each bird sees its neighbourhood only, it tries to alight its movement direction with the neighbouring birds. This means that if a predator attacks, once one bird changes its direction on the edge of a swarm, the impact cascades through the entire swarm. The key point here is that a bird in the middle or the other side of the swarm is not even aware of the predator. However, it constantly aligns its movement direction, which is good enough to avoid predators. Due to the existence of multiple neighbouring birds, the average alignment of all of them are considered. The second operator proposed by Raylond is separation which prevents collision between birds. Each bird avoids crowded areas. To avoid separation in the flock, each bird also updates its position to the average position of neighboring birds which is called cohesion. With these three simple local rule several global swarming behaviours can be achieved.

The interaction between individuals in a swarm can be done via environment too. For instance, ants in an ant colony leaves and sense pheromone when they move [3]. When they face multiple paths, they are more likely to choose the one with a higher amount of pheromone. For instance, if an ant finds a food source, it marks the path to the nest with pheromone. This attracts some ants towards the food who deposit pheromone to mark the path too. In case of multiple paths, the shortest one will be eventually used by all ants to bring the food source to the nest. This type of local interaction results in global intelligence without any centralized control unit.

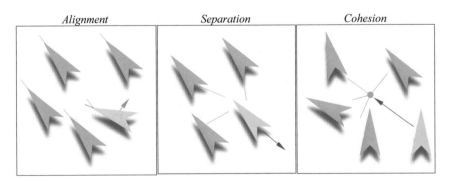

Fig. 1 Alignment, separation, and cohesion in a flock of birds

For more information about the Ant Colony Optimization algorithm, which was built based on these simple rules, please refer to chapter "Introduction to Nature-Inspired Algorithms".

Grasshopper Optimization Algorithm (GOA) [4] is a recent algorithm proposed by Saremi et al. that belongs to the family of Swarm Intelligence techniques. This algorithm mimics the navigation of adult grasshopper in nature when forming one of the largest swarms on the planet [5]. GOA has been widely used since its proposal. Some of the main advantages of this algorithm are: a small number of controlling parameters, adaptive exploratory and exploitative search pattern, and gradient-free mechanism. This chapter first dives into the main mechanisms of this algorithm. The controlling parameters, improvements, and applications of this algorithms are then discussed. The chapter also shows the application of this algorithm in the field of hand posture estimation, which is a key step in the overall process of hand gesture detection.

2 Grasshopper Optimization Algorithm

As discussed above, the GOA algorithm simulates the swarming behaviour of grasshoppers in nature. The mathematical equations and formulations proposed for this algorithm are given as follows [4]. In GOA, the position of the grasshoppers in the swarm represents a possible solution of a given optimization problem. The position of the i-th grasshopper is denoted as X_i and represented as given in Eq. 1.

$$X_i = S_i + G_i + A_i \tag{1}$$

where S_i is the social interaction, G_i is the gravity force on i-th grasshopper, and A_i shows the wind advection.

Equation 1 is made of three main components to simulate social interaction, impact of gravitational force, and wind advection. These three components simulate the movement of grasshoppers. To solve optimization problems, GOA only simulates the social interaction (S_i), which is defined as follows:

$$S_i = \sum_{j=1,\, j \neq i}^{N} s\left(d_{ij}\right) \widehat{d_{ij}} \tag{2}$$

$$d_{ij} = |x_j - x_i| \tag{3}$$

where d_{ij} is the Euclidean distance between the i-th and j-th grasshoppers. The $\widehat{d_{ij}} = \frac{x_j - x_i}{d_{ij}}$ is a unit vector from the i-th grasshopper to the j-th grasshopper.

In a grasshopper swarm, a grasshopper might face three forces depending on its location as compared to neighbouring grasshoppers: attraction, repulsion, and

Fig. 2 Comfort zone, attraction, and repulsion in the swarm of grasshoppers

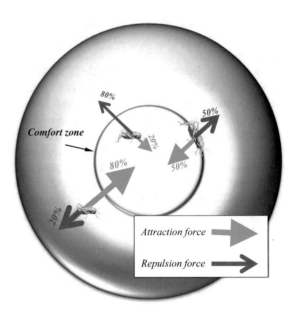

neutral. This means the space is divided into three area: before the comfort zone, inside the comfort zone, and after the comfort zone. To show the interaction between grasshoppers with respect to comfort area, Fig. 2 shows a conceptual model.

Saremi et al. used the following mathematical function to simulate these three zones and their forces:

$$s(r) = f e^{\frac{-r}{l}} - e^{-r} \tag{4}$$

where f indicates the intensity of attraction, and l is the attractive length scale.

The function s is illustrated in Fig. 3 to show how it impacts the social interaction (attraction and repulsion) of grasshoppers. This figure shows that the repulsion occurs from 0 to 2.079. This value can be normalized to any desired range when solving optimization problems. A grasshopper will be in a neutral position (no force applied) when the distance is equal to 2.079. As the grasshopper go further it faces more attractive force until about 5 where the magnitude of forces decreases due to the large distance. In GOA, it is assumed that that range of the function s is between 0 and 3.3.

In GOA, the function s has been designed with two controlling parameters: l and f as can be seen in Eq. 4. To see the changes in the shape of this function when changing the controlling parameters, Fig. 4 is given. This figure shows that the comfort, attractive, and repulsive zones can be changed significantly by tuning the controlling parameters. Despite the merits of the function s, it is not able to apply strong forces between grasshoppers with large distances between them. To resolve this issue, the distance between grasshoppers should be normalized (e.g. in [1, 4]).

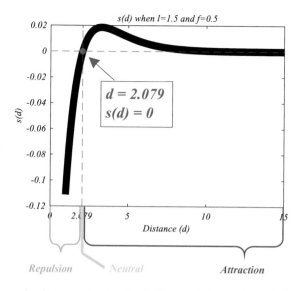

Fig. 3 The curve that is obtained using Eq. 4. The comfort zone (neutral situation) is where the distance is equal to 2.079. Repulsion occurs when $1 \leq d < 2.0790$. Attraction occurs when $2.0790 \leq d$

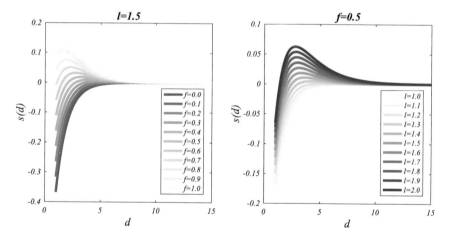

Fig. 4 Different curves can be obtained for the function s when changing its parameters

To solve optimization problems, a stochastic algorithm requires to perform exploration and exploitation effectively to find an accurate approximation of the global optimum. Saremi et al. equipped the above-mentioned equations with several parameters to show exploration and exploitation in different stages of optimization. Such modifications are required to converge to a solution at the end of optimization process. The mathematical model of position updating of grasshoppers in the GOA algorithm is as follows:

$$X_i^d = c \left(\sum_{j=1, \, j \neq i}^{N} c \frac{ub_d - lb_d}{2} s \left(|x_j^d - x_i^d| \right) \frac{x_j - x_i}{d_{ij}} \right) + \widehat{T_d} \qquad (5)$$

where ub_d is the upper bound in the d-th dimension, lb_d is the lower bound in the d-th dimension $s(r) = fe((-r)/l) - e(-r)$, (T_d) is the value of d-th dimension in the target (best solution found so far), and c is a decreasing coefficient to shrink the comfort area, repulsion area, and attraction area.

It should be noted that in Eq. 5, the S function is similar to the mathematical function discussed above. In Eq. 5, in addition, the inner c contributes to the reduction of the magnitude of repulsion/attraction forces between grasshoppers proportional to the number of iterations, while the outer c reduces the search coverage around the target as the iteration counter increases. The target is essential for the convergence of GOA algorithm, which is assumed to be the best solution obtained so far. This means this mathematical model requires the solutions to constantly converge and diverge towards the best solution obtained so far.

Saremi et al. argued that the parameter c should updated with the following equation to reduce exploration and increase exploitation proportional to the number of iteration as follows:

$$c = c_{Max} - l \frac{c_{Max} - c_{Min}}{L} \qquad (6)$$

where $cmax$ is the maximum value, $cmin$ is the minimum value, l indicates the current iteration, and L is the maximum number of iterations. In this chapter, we use 1 and 0.00001 values for $cmax$ and $cmin$, respectively.

3 Literature Review of Grasshopper Optimization Algorithm

As discussed in the introduction of this chapter, the GOA algorithm has been widely used in the literature. These works can be divided into two classes: those on the theory and improvement of GOA and those investigate the applications of GOA. The following subsections provide a brief literature review of the most recent works on this algorithm.

3.1 Different Variants of GOA

The original version of GOA is able to solve unconstrained single-objective optimization problems with continuous variables. Therefore, one might need to change its structure to solve problems of other types. In the literature, there have been several modifications to this algorithm to solve a wide range of problems.

To solve multi-objective problems, there are several works in the literature [6–9]. In [6, 7], an archive is employed to store and improve Pareto optimal solutions for a given optimization problem. Since the MOGOA algorithm maintains the multi-objective formulation of problems, it can be considered as a posteriori optimization algorithm. The authors also employed several mechanisms to improve the diversity of Pareto optimal solutions in the archive of MOGOA.

To solve binary and discrete problems, three works have attempted to modify GOA. Aljarah et al. [10] used a rounding technique to round continuous numbers in [0, 1] to their nearest integers. They, then used the binary values to select features. In [11, 12], four operators were employed to develop the binary version of GOA. They first modified the initialization of GOA. A local search, a percentile technique position updating, and a repair mechanism to fix infeasible solution were then integrated into the GOA algorithm.

To handle constraints, there are some works in the literate. The main published work on constrained GOA can be found in [13], in which a penalty function is used to penalize the grasshoppers that violate any of the constraints.

In the field of meta-heuristics, researchers often hybridize algorithms to improve the achieve a higher performance. This has been the case for the GOA algorithm too. In [14], Mafarja et al. equipped GOA with an evolutionary operator called evolutionary population dynamics to improve the exploration of this algorithm. In [15], the opposition-based version of GOA was proposed in which the authors used two stages. For one, the initialization of GOA was modified to generate random solutions and their opposites using opposition-based learning. For another, the position updating of GOA was integrated with opposition-based learning to create more diverse solutions in the population. To improve both exploration and exploitation of GOA, chaotic maps have been used in [16, 17] as well.

3.2 Applications of GOA

As mentioned above, the GOA algorithm is gradient-free, which has been the main reason of its popularity in different fields. This algorithm has been applied to the following problems:

- Trajectory optimization for multiple solar-powered UAVs [18, 19]
- Short-term load forecasting [20]
- Electrical characterisation of proton exchange membrane fuel cells stack [21]
- Data clustering [22]
- Scheduling of thermal System [23]
- Analyzing vibration signals from rotating machinery [24]
- Short-Term Wind Electric Power Forecasting [25]
- Optimal power flow [26]
- Optimal distribution system reconfiguration and distributed generation placement [27]

- Classification for biomedical purposes [28]
- Designing of linear antenna arrays [29]
- Automatic Seizure Detection in EEG Signals [30]
- Synthetic Aperture Radar Image Segmentation [31]
- Designing automatic voltage regulator system [32]
- Reconfiguration of partially shaded PV arrays [33]
- Training Neural Network [34]

4   Results

This section applies GOA on a set of benchmark problems and the problem of hand posture estimation.

4.1   Testing GOA on Some Benchmark Functions

In this subsection, a GOA with 10 solutions and 100 iterations is applied to several unimodal, multi-modal, and composite test functions. The test functions are commonly used in the field of optimization, and interested readers are referred to the original paper of GOA to find the details of the functions. For the results, several figures are given: history of sampled points, attraction/repulsion portion, trajectory of one variable, average objective value of all solutions, and the convergence curve. Note that the following code is used to updated c in this experiments:

Algorithm 1 Updating c for the experiments on the benchmark functions

---
if $t < T$ **then**
    $c = d\%2$
else
    $c = 2 + d\%2$
end if

Figures 5 and 6 show the results of GOA on unimodal, multi-modal, and composite test functions. Such test functions are linear without any local solution, which make them suitable for testing exploitation and convergence.

Inspecting the results in these figures, the following observations are made:

- The coverage of sampled point is high. This shows that the global search is high too. This is due to the mechanism in Algorithm 1, in which the first half of the iteration is used for exploration.

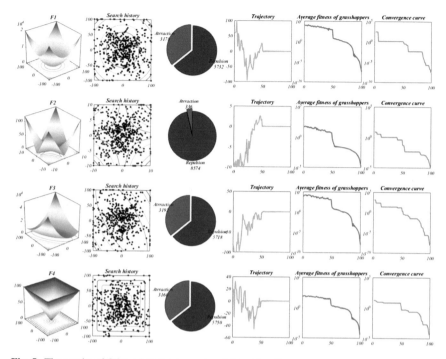

Fig. 5 The results of GOA when solving unimodal test functions

- The exploitation is high as well due to the high accuracy of the solution obtained as per the final solution obtained. Algorithm 1 requires solutions to exploit the target in the last 50% of iterations.
- The trajectory shows significant fluctuations in the first half of the iteration and gradual changes afterwards. Once again, this shows exploratory behavior that is substituted by exploitation after spending half of the iterations.
- The average fitness of all solutions decreases over the iterations. This shows that GOA is able to improve the quality of a random population iteratively.
- The convergence curve is accelerated proportional to the number of iterations, which shows that the exploitative behavior of GOA is increased over the course of iterations.
- Grasshoppers face different repulsive or attractive forces depending on the problem that they are solving. In most of test cases, the attraction is high due to the higher exploitation. However, GOA shows a large number of repulsions between grasshopper on some multi-modal test functions. This reduces the probability of local optima stagnation.

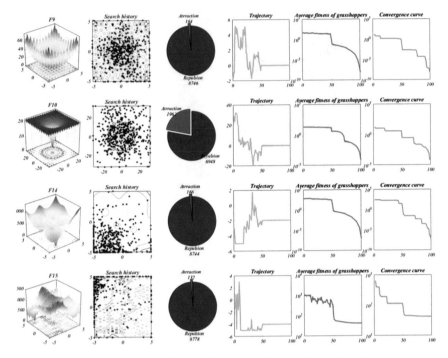

Fig. 6 The results of GOA when solving multi-modal and composite test functions

4.2 Applying GOA to the Problem of Hand Posture Estimation

This section investigates the application of the proposed GOA in the field of gesture detection. Gesture detection is one of the most popular topics in Human Computer Interaction (HCI). As its name implies, gesture detection refers to the process of detecting and interpreting humans gesture using computers. There are different types of gestures in each part of our bodies: body, head, face, hand, and so on. Needless to say, hand gesture recognition aims for detecting humans hand gestures. The general steps of a system for detecting hand gesture are illustrated in Fig. 7 [35].

The data from the hand can be acquired by gloves or cameras. The former method is accurate but requires to wear a device [36], which is not possible for some applications such as surgery. The latter method is vision-based [37] so the user interacts naturally in front of one or multiple cameras, but the accuracy and reliability is lower than gloves due to the resolution, occlusion, and noises. In the second phase, a model of hand is created in computer using the data acquired in the first phase. Again, there are many different approaches for modeling hand in computer: motion-based models, skeleton models, volumetric models, geometric models, etc. The third phase, feature extraction [38], extracts the essential elements for recognizing hand gesture in the last phase. The features that can be extracted highly depend on the model employed

Fig. 7 General steps of hand gesture recognition

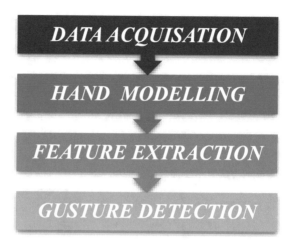

in the second phase. Some of the most important features are: number of fingers, orientation, location, angle of joints, shape, motion, etc. The last phase utilizes the extracted features in the third phase to recognize gestures. Basically, this phase is a classification phase where the extracted features are used to find the gesture in a repository. Popular tools in this phase are: Neural Networks (NNs) [39, 40], Support Vector Machine (SVM) [41, 42], etc.

Despite the importance of each phase, a gesture cannot be detected reliability and accurately without suitable hand modeling and feature extraction. The problem investigated in this subsection is to find the optimal values for a 3D model of hand based on the 2D image of hand acquired from one camera as shown in Fig. 8.

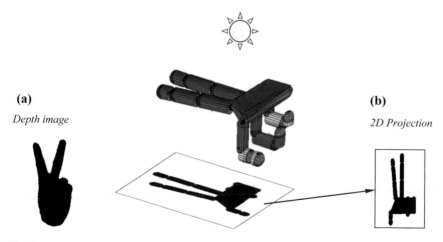

Fig. 8 a 2D binary image of hand **b** desired 3D model/2D projection with the least discrepancy with the 2D image

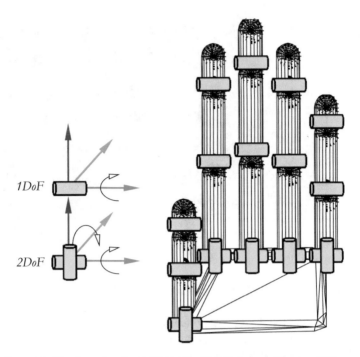

Fig. 9 3D model of hand employed with 20 (5 * 2 + 10 * 1) degree of freedom [43]

The 3D model employed in this work is a 20-Degree-of-Freedom (DoF) hand model [39]. The parameters are the angle of the joints as shown in Fig. 9. This figure shows that the joints attached to the palm are all of 2 DoF, whereas other joints have only 1 DoF.

With the above objective function and hand model, this problem can be formulated as follows:

$$\text{Minimize: } F(\overrightarrow{x}) \tag{7}$$

$$\overrightarrow{x} = [\theta_1, \theta_2, \theta_3, ..., \theta_{19}, \theta_{20}] \tag{8}$$

$$\text{Subject to: } 0 < \theta_1, \theta_2, \theta_3, ..., \theta_{19}, \theta_{20} < 1.8 \tag{9}$$

The F function calculates the discrepancy of the projection of the given 3D model of hand from the obtained image of hand. This problem is solved by 30 search agents over 300 iterations. The optimal values for the parameters and objective are presented in Table 1. The convergence curve, initial random model, and obtained optimal hand model are illustrated in Fig. 10.

The results show how well the GOA algorithm is able to improve the initial random solution to match the desired shape in a 3D space. The improvement is quite significant until roughly 60 iterations where the fitness function reaches 15000. After this point, the convergence curve faces gradual decrement. The run time of a gesture detection system is very important for real-time hand tracking. Therefore, we

Table 1 Best optimal values obtained for each angle in the hand model. Note that the best optimal objective value is: F = 13940

θ_1	θ_2	θ_3	θ_4	θ_5	θ_6	θ_7	θ_8	θ_9	θ_{10}
−0.001	−0.100	0.245	0.033	−0.001	−0.100	−0.100	−0.100	−0.049	0.368
θ_{11}	θ_{12}	θ_{13}	θ_{14}	θ_{15}	θ_{16}	θ_{17}	θ_{18}	θ_{19}	θ_{20}
0.931	0.902	−0.003	−0.100	−0.095	0.016	−0.006	0.069	−0.089	−0.002

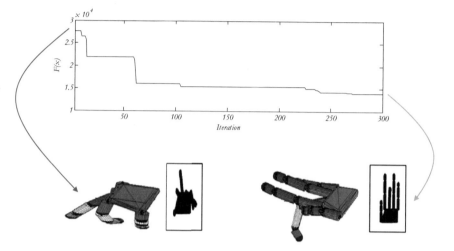

Fig. 10 Convergence curve, initial random model, and obtained optimal hand model

discuss the run time of the optimization process here as well. The whole optimization process took around 20 min on a machine with a dual core CPU and 4 GB RAM. The objective function evaluation is computationally the most expensive component, which is mainly because of its implementation in Matlab with lots of overheads in drawing and saving figures. It is worth noticing here that the actual convergence occurs around 60th iteration, so the run time can be reduced up to 5 min (20/4) by setting a threshold as the termination criterion. However, another approach is to reduce the number of grasshoppers, use a faster programming language, parallelizing the objective function evaluation, or using GPU to speed up the process. These considerations are outside the scope of this chapter but definitely have the potential to reduce the run time of the whole optimization process close to real-time.

5 Conclusion

This chapter first presented the mathematical model of the GOA algorithm. It then provided a brief literature review of this algorithm. It was discussed that this algorithm has been used in a large number of works and applied to a wide range of applications. The algorithm itself has been tuned, improved, and hybridized in the literature too. These show the merits of this algorithm is solving challenging problems. To experimentally observe the performance of GOA, several experiments were conducted. A set of test functions were solved by GOA, showing the ability of this algorithm in exploring and exploiting a search space. Then, the algorithm was applied to a challenging problem in the field of hand posture estimation.

The results of test functions shows that GOA is able to significantly improve the quality of the random population, improve the quality of the best solution obtains so far, show high exploration and exploitation, and fluctuate the variables of solution significantly in the first half of iterations to avoid local solutions. After applying GOA to the problem of hand posture estimation, it was observed that GOA benefits from steady convergence speed and it is reliable to find a reasonable solution for the problem investigated.

References

1. Blum, C., & Li, X. (2008). Swarm intelligence in optimization. In *Swarm intelligence* (pp. 43–85). Berlin: Springer.
2. Reynolds, C. W. (1987). Flocks, herds and schools: A distributed behavioral model. In *ACM SIGGRAPH computer graphics* (Vol. 21, No. 4, pp. 25–34). ACM.
3. Dorigo, M., & Birattari, M. (2011). Ant colony optimization. In *Encyclopedia of machine learning* (pp. 36–39). Boston: Springer.
4. Saremi, S., Mirjalili, S., & Lewis, A. (2017). Grasshopper optimisation algorithm: Theory and application. *Advances in Engineering Software, 105*, 30–47.
5. Topaz, C. M., Bernoff, A. J., Logan, S., & Toolson, W. (2008). A model for rolling swarms of locusts. *The European Physical Journal Special Topics, 157*(1), 93–109.
6. Mirjalili, S. Z., Mirjalili, S., Saremi, S., Faris, H., & Aljarah, I. (2018). Grasshopper optimization algorithm for multi-objective optimization problems. *Applied Intelligence, 48*(4), 805–820.
7. Tharwat, A., Houssein, E. H., Ahmed, M. M., Hassanien, A. E., & Gabel, T. (2017). MOGOA algorithm for constrained and unconstrained multi-objective optimization problems. *Applied Intelligence*, 1–16.
8. Lewis, A. (2009). LoCost: A spatial social network algorithm for multi-objective optimisation. In *IEEE Congress on Evolutionary Computation, 2009. CEC 2009* (pp. 2866–2870). IEEE.
9. Lewis, A. (2009). The effect of population density on the performance of a spatial social network algorithm for multi-objective optimisation. In *IEEE International Symposium on Parallel & Distributed Processing, 2009. IPDPS 2009.* (pp. 1–6). IEEE.
10. Aljarah, I., AlaM, A. Z., Faris, H., Hassonah, M. A., Mirjalili, S., & Saadeh, H. (2018). Simultaneous feature selection and support vector machine optimization using the grasshopper optimization algorithm. *Cognitive Computation*, 1–18.
11. Pinto, H., Pea, A., Valenzuela, M., & Fernndez, A. (2018). A binary grasshopper algorithm applied to the knapsack problem. In *Computer Science On-line Conference* (pp. 132–143). Springer, Cham.
12. Crawford, B., Soto, R., Pea, A., & Astorga, G. (2018). A binary grasshopper optimisation algorithm applied to the set covering problem. In *Computer Science On-line Conference* (pp. 1–12). Springer, Cham.
13. Neve, A. G., Kakandikar, G. M., & Kulkarni, O. (2017). Application of grasshopper optimization algorithm for constrained and unconstrained test functions. *International Journal of Swarm Intelligence and Evolutionary Computation, 6*(165), 2.

14. Mafarja, M., Aljarah, I., Heidari, A. A., Hammouri, A. I., Faris, H., AlaM, A. Z., et al. (2018). Evolutionary population dynamics and grasshopper optimization approaches for feature selection problems. *Knowledge-Based Systems, 145*, 25–45.
15. Ewees, A. A., Elaziz, M. A., & Houssein, E. H. (2018). Improved grasshopper optimization algorithm using opposition-based learning. *Expert Systems with Applications.*
16. Arora, S., & Anand, P. (2018). Chaotic grasshopper optimization algorithm for global optimization. *Neural Computing and Applications*, 1–21.
17. Saxena, A., Shekhawat, S., & Kumar, R. (2018). Application and development of enhanced chaotic grasshopper optimization algorithms. *Modelling and Simulation in Engineering.*
18. Wu, J., Wang, H., Li, N., Yao, P., Huang, Y., Su, Z., & Yu, Y. (2017). Distributed trajectory optimization for multiple solar-powered UAVs target tracking in urban environment by adaptive grasshopper optimization algorithm. *Aerospace Science and Technology, 70*, 497–510.
19. Wu, J., Wang, H., Li, N., Yao, P., Huang, Y., Su, Z., et al. (2017). Distributed trajectory optimization for multiple solar-powered UAVs target tracking in urban environment by adaptive grasshopper optimization algorithm. *Aerospace Science and Technology, 70*, 497–510.
20. Barman, M., Choudhury, N. D., & Sutradhar, S. (2018). A regional hybrid GOA-SVM model based on similar day approach for short-term load forecasting in Assam, India. *Energy, 145*, 710–720.
21. El-Fergany, A. A. (2017). Electrical characterisation of proton exchange membrane fuel cells stack using grasshopper optimiser. *IET Renewable Power Generation, 12*(1), 9–17.
22. ukasik, S., Kowalski, P. A., Charytanowicz, M., & Kulczycki, P. (2017). Data clustering with grasshopper optimization algorithm. In *Federated Conference on Computer Science and Information Systems (FedCSIS), 2017* (pp. 71–74). IEEE.
23. Rajput, N., Chaudhary, V., Dubey, H. M., & Pandit, M. (2017). Optimal generation scheduling of thermal System using biologically inspired grasshopper algorithm. In *2nd International Conference on Telecommunication and Networks (TEL-NET), 2017* (pp. 1–6). IEEE.
24. Zhang, X., Miao, Q., Zhang, H., & Wang, L. (2018). A parameter-adaptive VMD method based on grasshopper optimization algorithm to analyze vibration signals from rotating machinery. *Mechanical Systems and Signal Processing, 108*, 58–72.
25. Zhao, H., Zhao, H., & Guo, S. (2018). Short-term wind electric power forecasting using a novel multi-stage intelligent algorithm. *Sustainability, 10*(3), 881.
26. Buch, H., & Trivedi, I. N. On the efficiency of metaheuristics for solving the optimal power flow. *Neural Computing and Applications*, 1–19.
27. Ahanch, M., Asasi, M. S., & Amiri, M. S. (2017). A grasshopper optimization algorithm to solve optimal distribution system reconfiguration and distributed generation placement problem. In *IEEE 4th International Conference on Knowledge-Based Engineering and Innovation (KBEI), 2017* (pp. 0659–0666). IEEE.
28. Ibrahim, H. T., Mazher, W. J., Ucan, O. N., & Bayat, O. A grasshopper optimizer approach for feature selection and optimizing SVM parameters utilizing real biomedical data sets. *Neural Computing and Applications*, 1–10.
29. Amaireh, A. A., Alzoubi, A., & Dib, N. I. (2017). Design of linear antenna arrays using antlion and grasshopper optimization algorithms. In *IEEE Jordan Conference on Applied Electrical Engineering and Computing Technologies (AEECT), 2017* (pp. 1–6). IEEE.
30. Hamad, A., Houssein, E. H., Hassanien, A. E., & Fahmy, A. A. (2018). Hybrid grasshopper optimization algorithm and support vector machines for automatic seizure detection in EEG signals. In *International Conference on Advanced Machine Learning Technologies and Applications* (pp. 82–91). Springer, Cham.
31. Sharma, A., & Sharma, M. (2017). SAR image segmentation using grasshopper optimization algorithm.
32. Hekimolu, B., & Ekinci, S. (2018). Grasshopper optimization algorithm for automatic voltage regulator system. In *2018 5th International Conference on Electrical and Electronic Engineering (ICEEE)* (pp. 152–156). IEEE.
33. Fathy, A. (2018). Recent meta-heuristic grasshopper optimization algorithm for optimal reconfiguration of partially shaded PV array. *Solar Energy, 171*, 638–651.

34. Heidari, A. A., Faris, H., Aljarah, I., & Mirjalili, S. (2018). An efficient hybrid multilayer perceptron neural network with grasshopper optimization. *Soft Computing*, 1–18.
35. Bansal, B. (2016). Gesture recognition: A survey. *International Journal of Computer Applications*, *139*(2).
36. Smith, A. V. W., Sutherland, A. I., Lemoine, A., & Mcgrath, S. (2000). U.S. Patent No. 6,128,003. Washington, DC: U.S. Patent and Trademark Office.
37. Garg, P., Aggarwal, N., & Sofat, S. (2009). Vision based hand gesture recognition. *World Academy of Science, Engineering and Technology*, *49*(1), 972–977.
38. Yang, M. H., Ahuja, N., & Tabb, M. (2002). Extraction of 2d motion trajectories and its application to hand gesture recognition. *IEEE Transactions on pattern analysis and machine intelligence*, *24*(8), 1061–1074.
39. Murakami, K., & Taguchi, H. (1991). Gesture recognition using recurrent neural networks. In *Proceedings of the SIGCHI Conference on Human factors in Computing Systems* (pp. 237–242). ACM.
40. Stergiopoulou, E., & Papamarkos, N. (2009). Hand gesture recognition using a neural network shape fitting technique. *Engineering Applications of Artificial Intelligence*, *22*(8), 1141–1158.
41. Dardas, N. H., & Georganas, N. D. (2011). Real-time hand gesture detection and recognition using bag-of-features and support vector machine techniques. *IEEE Transactions on Instrumentation and Measurement*, *60*(11), 3592–3607.
42. Saha, S., Konar, A., & Roy, J. (2015). Single person hand gesture recognition using support vector machine. In *Computational advancement in communication circuits and systems* (pp. 161–167). Springer, New Delhi.
43. Malvezzi, M., Gioioso, G., Salvietti, G., Prattichizzo, D., & Bicchi, A. (2013). SynGrasp: A matlab toolbox for grasp analysis of human and robotic hands. In *IEEE International Conference on Robotics and Automation (ICRA), 2013* (pp. 1088–1093). IEEE.

Multi-verse Optimizer: Theory, Literature Review, and Application in Data Clustering

Ibrahim Aljarah, Majdi Mafarja, Ali Asghar Heidari, Hossam Faris and Seyedali Mirjalili

Abstract Multi-verse optimizer (MVO) is considered one of the recent metaheuristics. MVO algorithm is inspired from the theory of multi-verse in astrophysics. This chapter discusses the theoretical foundation, operations, and main strengths behind this algorithm. Moreover, a detailed literature review is conducted to discuss several variants of the MVO algorithm. In addition, the main applications of MVO are also thoroughly described. The chapter also investigates the application of the MVO algorithm in tackling data clustering tasks. The proposed algorithm is benchmarked by several datasets, qualitatively and quantitatively. The experimental results show that the proposed MVO-based clustering algorithm outperforms several similar algorithms such as Particle Swarm Optimization (PSO), Genetic Algorithm (GA), and Dragonfly Algorithm (DA) in terms of clustering purity, clustering homogeneity, and clustering completeness.

I. Aljarah · H. Faris
King Abdullah II School for Information Technology, The University of Jordan,
Amman, Jordan
e-mail: i.aljarah@ju.edu.jo

H. Faris
e-mail: hossam.faris@ju.edu.jo

M. Mafarja
Department of Computer Science, Faculty of Engineering and Technology,
Birzeit University, PoBox 14 Birzeit, Palestine
e-mail: mmafarja@birzeit.edu

A. A. Heidari
School of Surveying and Geospatial Engineering, University of Tehran, Tehran, Iran
e-mail: as_heidari@ut.ac.ir

S. Mirjalili (✉)
Institute of Integrated and Intelligent Systems, Griffith University, Nathan,
Brisbane, QLD 4111, Australia
e-mail: seyedali.mirjalili@griffithuni.edu.au

© Springer Nature Switzerland AG 2020
S. Mirjalili et al. (eds.), *Nature-Inspired Optimizers*, Studies in Computational
Intelligence 811, https://doi.org/10.1007/978-3-030-12127-3_8

123

Keywords Optimization · Meta-heuristics Multi-verse optimizer ·
Swarm intelligence · MVO · Data clustering

1 Introduction

Swarm intelligence algorithms imitate the systems in nature such as bird flocks,
ant colonies, and fish schools. The behavior of these system members is based on
the communication ways between the individuals in that system in order to find the
best food sources. Many researchers tend to solve different optimization problems
by employing these types of methods, such as Ant Colony Optimization (ACO)
[68], Particle Swarm Optimization (PSO) [58], Deferential Evolution (DE) [19],
Glowworm Swarm Optimization (GSO) [9], and many others [12, 22, 30, 38, 42,
54]. Swarm intelligence algorithms have been used in several applications such as
global function optimization [11, 27, 39–41, 43], optimizing neural networks [1, 5,
6, 23, 26], clustering analysis [2, 7, 9, 69], machine learning [4, 32, 53, 55–57], spam
and intrusion detection systems [8, 10, 24, 29], link prediction [14], software effort
estimation [34], and bio-informatics [13, 18]. The main objective of the majority
of the swarm intelligence algorithms is to locate the global solution for the given
optimization problem.

Clustering analysis is an unsupervised learning technique to organize of unlabeled
data into similar groups called clusters [45]. The clustering algorithm combines the
similar data objects having common features and splits them into different groups
based on a proximity metric. Clustering is in contrast with supervised learning, such
as classification and regression, where the data labels are available. Partitional clustering methods are the most popular clustering approaches (e.g. k-Means algorithm) that
employed in the literature [17]. In partitional clustering, data objects are divided into
predefined clusters using some proximity measures. This type of clustering methods
have been applied to diverse fields including: document categorization [16], image
segmentation [62], and network intrusion detection [35]. Partitional clustering methods have some characteristics such as simple to implement and have low computation
complexity. However, they suffer from some weaknesses such as the premature convergence possibility and very sensitive to the initial centroids. Therefore, many new
algorithms have been developed in literature to overcome such drawbacks. Some
of these new methods benefit from the capabilities of nature-inspired and swarm
intelligence algorithms.

Multi-verse optimizer (MVO) is one of the recent swarm intelligence algorithms,
which is inspired from the theory of multi-verse that discuss how the big bangs create
multiple universes and how these universes interact with each other through different
types of holes. The objective of the MVO algorithm is to locate the global solution
for the given optimization problem. MVO has been used in several applications such
as training feed-forward neural networks [25], optimizing support vectors machines
[31], engineering applications [61], power systems optimization [50].

In this chapter, the theoretical foundation, operations, and main strengths behind MVO algorithm is discussed. Then a detailed literature review is conducted to discuss several variants of the MVO algorithm. Moreover, we applied the MVO algorithm on the clustering analysis problems, which takes into account the advantages of the efficiency of the MVO to locate optimal solutions. In addition, the proposed clustering algorithm is tested on real data sets with different characteristics to evaluate its performance.

The rest of the chapter is organized as follows: the overview of MVO algorithm is presented in Sect. 2. Section 3 provides a detailed literature review of MVO. The design of the proposed clustering algorithm based on MVO is presented in Sect. 4. Section 5 shows the implementation and results of the MVO clustering approach. Finally, Sect. 6 shows some final remarks and future works.

2 Multi-verse Optimizer

Multi-verse optimization algorithm (MVO) is one of the recent swarm intelligence methods, which was proposed by Mirjalili et al. [61]. MVO algorithm is inspired from the theory of multi-verse in astrophysics. The multi-verse theory clarifies how the big bangs create multiple universes and how these universes interact with each other through different types of holes such as white, black, and worm holes. The black and white interacts though a tunnel, which represents a transfer between two universe pairs. Black holes attract everything and white holes emit everything. On the other hand, a worm hole creates a tunnel though time and connects different parts of a universe. Every universe has an inflation rate that causes its expansion through space. In MVO algorithm, the authors in [61] utilize the concepts of white hole and black hole in order to explore the wormholes to exploit the search spaces to formulate a population-based algorithm. They assumed that each solution represents a universe and each variable/attribute in the solution represents an object in that universe. In addition, each solution has a fitness value (inflation rate) to reflect the solution quality, which is calculated by the corresponding objective function.

A good objective value is given to a solution when white holes are appeared, while a worse objective value is given to the the the solution if the black holes are appeared. With more interactions between white and black holes, the variable values of the good solutions are moved to poor solutions.

The main mathematical model of MVO algorithm depends on Eqs. (1) and (2) which are described as follows:

$$
X_i^j = \begin{cases} X_k^j, & r_1 < NI(U_i) \\ X_i^j, & r_1 \geq NI(U_i) \end{cases} \tag{1}
$$

where X_i^j represents the jth object of the ith universe, r_1 is a random number in the range (0,1), $NI(U_i)$ represents the normalized inflation rate of the ith universe and X_k^j represents the jth object of the kth universe.

$$X_i^j = \begin{cases} \begin{cases} ((X_j + TDR \times (ub - lb) \times r_4 + lb), & r_3 < 0.5 \\ (X_j - TDR \times (ub - lb) \times r_4 + lb), & r_3 \geq 0.5 \end{cases}, & r_2 < WEP \\ X_i^j, & r_2 \geq WEP \end{cases} \quad (2)$$

where X_j represents the jth parameter of the best universe obtained so far, ub is the upper bound, lb is the lower bound, Traveling Distance Rate (TDR) is a coefficient, Wormhole Existence Probability (WEP) is also a coefficient, r_2, r_3 and r_4 are random numbers in the range (0, 1). WEP and TDR represent adaptive variables, which WEP is employed to enhance exploitation, and TDR is employed to enhance exploitation around the best solution obtained so far. The adaptive formula for WEP, and TDR coefficients are used as follows:

$$WEP = min + l \times (\frac{max - min}{L}) \quad (3)$$

$$TDR = 1 - \frac{l^{1/p}}{L^{1/p}} \quad (4)$$

where p represents the exploitation factor. Where min is the minimum and max is the maximum, l is the current iteration, and L shows the maximum iterations.

In the MVO algorithm, the optimization mechanism begins with initializing a set of universes with random numbers. At each iteration, variables in the universes with high fitness values (high inflation) move toward the universes with low fitness values via white/black holes. Meanwhile, every universe encounters random theoretical transfer in its variables through wormholes towards the best universe. This process is iterated until a pre-defined maximum number of iterations. In addition, the MVO algorithm saves the best solution during optimization and utilizes it to impact on the rest of solutions. Pseudocode of MVO is shown in Algorithm 1.

3 Literature Review

Many researchers have been modified the original MVO algorithm and applied it in different applications. In this section, the recent versions of MVO and their applications are thoroughly described. The research works about MVO can be divided into three main categories: MVO variants, MVO applications, and MVO implementations.

Algorithm 1 Pseudo-code of MVO algorithm

Input: Population size and number of iterations (L).
Output: The best universe and its inflation rate.
Define: SU = Sorted universes, NI = Normalized inflation rate, Black_hole_index = i, r_1, r_2, r_3, r_4 = rand([0,1])
Initialize all random universes x_i ($i = 1, 2, \ldots, n$), WEP, TDR, and best universe.
while (end condition is not met) **do**
    Calculate the fitness of universes.
    for (each $Universe_i$) **do**
        Update WEP and TDR
        for (each $Object_j$) **do**
            if $r_1 < NI(U_i)$ **then**
                White_hole_index = RouletteWheelSelection($-NI$)
                U(Black_hole_index, j) = SU(White_hole_index, j)
            if $r_2 < WEP$ **then**
                if $r_3 < 0.5$ **then**
                    $U(i, j) = $ Best_universe(j) $+ TDR \times ((ub(j) - lb(j)) \times r_4 + lb(j))$;
                else
                    $U(i, j) = $ Best_universe(j) $- TDR \times ((ub(j) - lb(j)) \times r_4 + lb(j))$;
Return: The best universe

3.1 MVO Variants

In [60], the multi-objective version of MVO (MOMVO) was proposed to solve multi-objective problems. MOMVO utilizes the same way in the original MVO algorithm to update the position of solutions in a multi-objective search space to obtain Pareto optimal solutions. The MOMVO is compared with other multi-objective algorithms using popular benchmarks based on a variety of evaluation measures. The results showed that the MOMVO algorithm is able to locate Pareto optimal fronts for most benchmarks that used in the experiments.

Another variant of MVO algorithm in [44] is proposed. The authors in this research work combined the original MVO algorithm with random walk mechanism called levy flight to solve numerical and Network-on-Chip (NoC) scheduling problems. The Levy flight that integrated in best universe helps the MVO algorithm to escape from the local optimum. Experimental results of modified algorithm with well-known benchmark and scheduling benchmarks show good results in terms of solutions quality and convergence speed.

In [21, 67], chaotic maps are used to enhance the efficiency of MVO algorithm. The authors use one-dimensional, non-invertible maps to generate a set of chaotic values to tune the MVO's parameters. The proposed algorithm in [21] was applied to solve the feature selection problem, while authors in [67] applied chaotic-based MVO algorithm to tackle the engineering optimization problems.

Stud selection and crossover (SSC) operator is used in [59] to improve the efficiency of the original MVO algorithm. In this work, the best universe is utilized to provide extra information for other universes in the search space using general genetic operators. The experimental results showed that the added modification into MVO can enhance the MVO's capabilities to achieve the optimal solutions.

Another version of MVO is proposed in [63]. The authors improved the MVO algorithm by presenting a new concept to MVO optimization process called exponential function inflation size. Furthermore, new control parameters are introduced to improve the effectiveness of the MVO algorithm. The results of the proposed work successfully proved increasing of the MVO's convergence rate. In [66], the quantum theory and its representation was used to modify the MVO algorithm to enhance the capabilities of MVO in locating the optimal solution. The proposed algorithm showed superiority in solving numerical optimization problems compared to the original MVO. A binary version of MVO was proposed in [73]. The authors used a percentile binary operator to formulate a binary MVO algorithm. The algorithm was applied and evaluated using set covering problem (SCP).

3.2   MVO Applications

Due to the unique properties of MVO, many research applications from different domains have been employed the MVO algorithm. Majority of applications which applied the MVO are related to the data mining, machine learning, and engineering-related applications. In the following paragraphs, an overview of these applications is provided.

The authors in [25, 37] employed MVO for training the Multi-layer Perceptron (MLP) neural network. The MVO algorithm is used to locate the best set of MLP's weights that maximize the classification accuracy. The MVO was tested using different well-known datasets, and its results were promising compared with other popular algorithms. In [15], the authors combined the MVO algorithm with an artificial neural network (ANN) to develop intrusion detection system. The MVO is used to train ANN to distinguish between the normal and abnormal connections. Moreover, the MVO algorithm was used in [31] to optimize the Support Vector Machines (SVMs)'s parameters. In addition, the authors developed a robust approach based on MVO algorithm for selecting optimal features. The results showed that MVO can obtain a high classification accuracy.

In [69], the search capabilities of MVO was utilized to realize clustering problems. The authors benefited from the powerful operators of the MVO algorithm to formulate MVO as clustering approaches in order to avoid the drawbacks of the partitional clustering algorithms. The clustering-based MVO approaches were evaluated based on well-known datasets and showed high quality results. In addition, an automatic clustering algorithm based on MVO algorithm in [20] was utilized to formulate the binarization process of images in historical handwritten documents.

3.2.1   Engineering Applications

MVO has been applied to a wide variety of engineering applications such as design and tuning controllers, power dispatch problems, power consumption and many others as described in below.

The authors in [33, 48] used MVO for optimizing the fuzzy-PID controller in two-area power system with HVDC link connection, and identifying the optimal parameters of PEMFC model. Their experimental results showed the effectiveness of MVO in comparison with other optimizers. In [47], the authors applied the MVO to optimize fuzzy-PID controller and improve power system's frequency regulation. Moreover, the authors in [49] investigated the effects of TCPS, SMES, and DFIG on load frequency control by optimizing the fuzzy-PID controller parameters using MVO algorithm. Ali et al. in [3] applied the MVO for parameter extraction of photovoltaic generating units.

In [71], the economic load dispatch (ELD), which is a class of non-convex and highly nonlinear constrained optimization problems, was optimized by MVO algorithm. Results showed the effectiveness of MVO in this type of problems. In [75–77], MVO was used to deal with the energy consumption estimation and load forecasting problem. The role of MVO algorithm in these researches is to predict the peak load and natural gas consumption. Another application that applied MVO algorithm to find the best solution of load frequency control problem in power system was proposed in [36]. Other MVO-based approaches in power engineering problems include [46, 72].

3.3 MVO Implementations

Mirjalili et al. in [61] released a demo Matlab version[1] of MVO beside the original paper of MVO. Another implementation of MVO was presented by Vivek et al. [74]. The authors described their implementation of MVO as a toolkit in LabVIEW, which is an integrated development environment for designing measurements, tests and control systems and applications. Furthermore, Faris et al. implemented MVO as part of EvoloPy[2] [28], which is an open source optimization framework written in Python.

4 Application of MVO in Solving Data Clustering Problems

The proposed algorithm is based on MVO and consists of four main stages: initialization stage, fitness evaluation, and centroids update.

In the initialization phase, first, a population P with n universes $U_1, U_2,..., U_n$ is initialized using uniform randomization within the given search space. Each universe U_i forms a clustering solution, which represents a vector of sub-vectors/centroids such as $U_i = \{c_{i1}, c_{i2},...,c_{ik}\}$, where each centroid is represented with m-dimensions vector such as $c_{ij} = (c_{ij1}, c_{ij2},...,c_{ijm})$, $j = 1, 2, \ldots, k$, and k represents the prede-

---

[1]http://www.alimirjalili.com/MVO.html.
[2]https://github.com/7ossam81/EvoloPy.

fined number of the clusters/centroids. Each centroid represents a sub-vector that consists of m dimensions, which reflects the number of the features in the dataset D and belongs to a set of k clusters such as $\mathscr{C} = \{C_1, C_2, \ldots, C_k\}$. The length of each universe $|U_i| = k \times m$. All universes are initialized randomly such that, each universe is initialized by k data points, which are selected from the given dataset.

Then, the inflation rate $(I(U_i))$ is calculated using some fitness/objective function to evaluate the clustering quality. In this chapter, we used the Sum of Squared Errors (SSE) [51, 64]. The SSE fitness function is calculated based on the Euclidean distances between the data points in the dataset and the nearest centroids in the clusters. The SSE can be calculated using the following formula:

$$SSE = \sum_{j=1}^{k} \sum_{i=1}^{|U_i|} \sigma(c_j, r_i)^2 \tag{5}$$

where σ represents the Euclidean distance between the centroid c_j and ith data point r_i, which is represented with m-dimensions such as $r_i = (r_{i1}, r_{i2}, \ldots, r_{im})$, and is given by the following equation:

$$\sigma(c_j, r_i) = \sqrt{\sum_{w=1}^{m} (c_{jw} - r_{iw})^2} \tag{6}$$

where r_{iw} represents wth dimension of the ith data point that belongs to the cluster with c_j centroid.

The fitness function SSE is utilized here to evaluate the goodness of each universe, such as each universe tries to minimize its fitness.

Through the process of MVO-based clustering algorithm, each universe is updated its centroids using MVO updating equations (Eq. 1 and Eq using the best universe 2). Then, each universe is moved toward the best universe by updating its centroids. After that, the SSE fitness function is re-evaluated. This iterative process is continued until the maximum number of iterations is achieved. After the clustering process is terminated, the best universe (clustering solution) is utilized to evaluate the clustering quality. Figure 1 depicts the encoding scheme of the universes.

The overall flowchart of the MVO-based clustering technique is shown in Fig. 2.

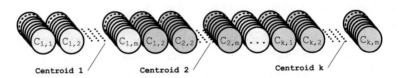

Fig. 1 Encoding scheme

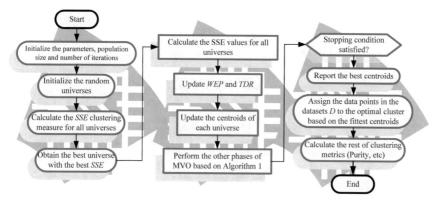

Fig. 2 Flowchart of MVO-based clustering technique

5 Implementation and Results

In this section, the proposed MVO-based clustering is applied to eight benchmark datasets with various number features and instances, which are obtained from UCI Machine Learning Repository [52]. Table 1 shows the details of the utilized benchmarks.

All the experiments and comparative evaluations in this research are performed and organized using a PC with Intel Core(TM) i5-2400 3.1GHz CPU and 4.0GB RAM. All tests are implemented under the same fair computational conditions.

The parameters of utilized techniques is tabulated in Table 2. The maximum number of iterations for optimizers is set to 200. The population size is set to 50 and the results are collected over 10 independent runs.

Table 1 List of utilized datasets

Dataset	Features	Instances	Classes
Iris	4	150	3
Breast cancer	10	683	2
Ecoli			
Seeds	7	210	3
Australian	15	653	2
Glass	9	214	6
Habitit			
Heart	13	270	2

Table 2 Parameters of algorithms

Algorithm	Parameter	Value
PSO	Inertia factor	0.1
	c_1	2
	c_2	2
DE	C Crossover	0.5
	F Scaling	[0.2 0.8]
GA	Crossover	0.8
	Mutation	0.02
MVO	WEP_{max}	1
	WEP_{min}	0.2

5.1 Evaluation Measures

To evaluate the clustering quality of the final solutions, we used 6 popular clustering measures: SSE, Homogeneity, Completeness, VMeasure, and Purity [9, 65, 70].

- SSE: which represents the fitness function of the optimization algorithm and given in the previous section in Eq. (6).
- Homogeneity score: which represents a score when each cluster contains only members of a single actual cluster, and given in Eq. (7)

$$Homogeneity = 1 - \frac{H(C|L)}{H(C)} \tag{7}$$

- Completeness score: which represents a score, when all members of a given actual cluster are assigned to the same cluster, and given in Eq. (8):

$$Completeness = 1 - \frac{H(L|C)}{H(L)} \tag{8}$$

where $H(C|L)$ is the conditional entropy of the actual clusters given the cluster assignments and is given by:

$$H(C|L) = -\sum_{i=1}^{k}\sum_{j=1}^{q}\frac{n_{i,j}}{R} \cdot \log\left(\frac{n_{i,j}}{n_j}\right) \tag{9}$$

where q is the number of actual clusters in the data set; k is the number of clusters that is generated from the clustering process; R is the total number of data points; n_i and n_j are the number of data points respectively belonging to actual cluster i and output cluster j, and finally $n_{i,j}$ the number of data points from actual cluster

C_i assigned to output cluster L_i. H(C) represents the entropy of the actual clusters and is given by:

$$H(C) = -\sum_{i=1}^{k} \frac{n_i}{R} \cdot \log\left(\frac{n_i}{R}\right) \tag{10}$$

The conditional entropy of clusters given class $H(L|C)$ and the entropy of clusters $H(L)$ are defined in a symmetric manner.

- VMeasure: is the harmonic mean that represents the combination of the Homogeneity and Completeness scores. VMeasure is given in Eq. (11):

$$V Measure = 2 \cdot \frac{Homogeneity \cdot Completeness}{Homogeneity + Completeness} \tag{11}$$

- Purity: is the percentage of data points that are clustered correctly. Purity is calculated using Eq. (12):

$$Purity = \frac{1}{R} \sum_{i=1}^{k} n_{i,j} \tag{12}$$

5.2 Results and Analysis

This section is devoted to reporting the attained results and discussion on the observations. Table 3 tabulates the recorded SSE results of MVO, PSO, DE, and GA in treating all test cases. As per results on Table 3, it is seen that the MVO can detect the relatively best clusters with the minimum *SSE* results for all datasets. Table 4 compares the clustering homogeneity results of MVO, PSO, DE, and GA in solving all cases. As per results in Table 4, we see that MVO can provide superior homogeneity results with acceptable Stdev values compared to other peers on 87.5% of datasets. Table 5 compares the performance of MVO based on the clustering completeness results versus other peers on all benchmark datasets. Based on results in Table 5, we vividly observe that both MVO and PSO have attained the highest completeness measures compared to DE and GA methods. The MVO is superior on Breast, Glass, Habitit, and Heart datasets, while PSO obtains higher completeness degrees on Iris, Ecoli, Seeds, and Australian cases. This fact indicates that, in the case of MVO and PSO, more members of a given actual cluster are assigned to the same cluster compared to those in GA-based and DE-based clustering techniques.

Table 6 compares the efficacy of MVO based on the clustering Vmeasure results versus other optimizers in dealing with all datasets. Based on this measure, MVO can significantly outperform all other techniques on all datasets. Table 7 shows the clustering purity results of MVO versus those of other optimizers in tackling all problems. From Table 7, it is detected that MVO can outperform GA, PSO, and DE on Iris, Breast, Ecoli, Seeds, Australian, Glass, Habitit, and Heart test cases. Hence, it can demonstrate a superior performance on 75% of datasets. In addition, even with

Table 3 Comparison of *SSE* results

Dataset	Measure	DE	PSO	GA	MVO
Iris	Avg	10.08043	10.33011	16.87869	**6.99900**
	Stdev	0.84519	1.74583	2.52121	0.00046
Breast	Avg	250.90482	350.28694	432.21967	**205.91133**
	Stdev	12.94197	44.31318	85.10503	0.00959
Ecoli	Avg	39.23229	39.95720	59.53466	**23.74054**
	Stdev	5.41344	10.82696	10.39184	5.77495
Seeds	Avg	38.75960	31.95170	55.00826	**25.54383**
	Stdev	4.34617	5.63711	9.31363	5.71475
Australian	Avg	860.05042	863.36608	1141.46918	**725.42315**
	Stdev	70.50186	48.10578	96.89295	99.05781
Glass	Avg	49.24797	46.52517	89.23814	**29.13085**
	Stdev	5.87662	9.06443	13.29072	2.56418
Habitit	Avg	137.08812	143.24364	180.64517	**120.80211**
	Stdev	3.76099	14.22743	20.99287	0.00543
Heart	Avg	366.10345	376.74819	425.62577	**314.06940**
	Stdev	19.14247	29.49627	24.33011	5.06730

Table 4 Comparison of clustering homogeneity results

Dataset	Measure	DE	PSO	GA	MVO
Iris	Avg	0.72778	0.65750	0.60002	**0.73642**
	Stdev	0.04379	0.07052	0.09578	0.00000
Breast	Avg	0.60761	0.54630	0.54180	**0.71033**
	Stdev	0.07130	0.09611	0.10294	0.00288
Ecoli	Avg	0.43868	0.22629	0.44054	**0.50214**
	Stdev	0.10838	0.14762	0.08253	0.13705
Seeds	Avg	0.55015	0.54263	0.54015	**0.61098**
	Stdev	0.10567	0.11405	0.06536	0.09793
Australian	Avg	0.14170	0.11327	0.08651	**0.23692**
	Stdev	0.15328	0.16348	0.10992	0.20784
Glass	Avg	0.18996	0.17044	**0.24416**	0.24341
	Stdev	0.05362	0.07987	0.04901	0.03544
Habitit	Avg	0.21828	0.17309	0.13965	**0.23047**
	Stdev	0.05856	0.08066	0.05032	0.00000
Heart	Avg	0.13902	0.17086	0.14584	**0.25875**
	Stdev	0.11303	0.11114	0.09743	0.06571

Table 5 Comparison of clustering completeness results

Dataset	Measure	DE	PSO	GA	MVO
Iris	Avg	0.75507	**0.82877**	0.69056	0.74749
	Stdev	0.04469	0.09641	0.09905	0.00000
Breast	Avg	0.64558	0.60237	0.59705	**0.71908**
	Stdev	0.05207	0.07163	0.06791	0.00244
Ecoli	Avg	0.56485	**0.74693**	0.52583	0.71637
	Stdev	0.12341	0.17063	0.08403	0.04119
Seeds	Avg	0.64305	**0.68222**	0.61663	0.67855
	Stdev	0.03752	0.05097	0.05254	0.03824
Australian	Avg	0.16857	**0.54091**	0.09380	0.35883
	Stdev	0.13989	0.40827	0.11331	0.27707
Glass	Avg	0.46231	0.46871	0.40213	**0.50376**
	Stdev	0.08873	0.11835	0.08900	0.07557
Habitit	Avg	0.17518	0.14604	0.11802	**0.18044**
	Stdev	0.05215	0.06358	0.05023	0.00000
Heart	Avg	0.13881	0.20987	0.15514	**0.25761**
	Stdev	0.11186	0.08492	0.10498	0.06283

Table 6 Comparison of clustering VMeasure results

Dataset	Measure	DE	PSO	GA	MVO
Iris	Avg	0.74096	0.72629	0.64046	**0.74191**
	Stdev	0.04231	0.02481	0.09045	0.00000
Breast	Avg	0.62582	0.57256	0.56719	**0.71468**
	Stdev	0.06228	0.08503	0.08781	0.00266
Ecoli	Avg	0.49188	0.31740	0.47512	**0.58060**
	Stdev	0.11438	0.20040	0.06779	0.10298
Seeds	Avg	0.58691	0.59593	0.57184	**0.63709**
	Stdev	0.06628	0.06504	0.03513	0.05412
Australian	Avg	0.14957	0.11904	0.08964	**0.23970**
	Stdev	0.14951	0.16396	0.11133	0.20352
Glass	Avg	0.26717	0.24495	0.30203	**0.32666**
	Stdev	0.06913	0.10986	0.05786	0.04368
Habitit	Avg	0.19425	0.15791	0.12762	**0.20241**
	Stdev	0.05507	0.07142	0.05026	0.00000
Heart	Avg	0.13890	0.18129	0.15021	**0.25816**
	Stdev	0.11245	0.11056	0.10090	0.06432

Table 7 Comparison of clustering purity results

Dataset	Measure	DE	PSO	GA	MVO
Iris	Avg	0.86733	0.77133	0.75333	**0.88667**
	Stdev	0.03777	0.09270	0.08433	0.00000
Breast	Avg	0.92718	0.90773	0.90587	**0.95393**
	Stdev	0.01976	0.02804	0.03512	0.00060
Ecoli	Avg	0.69235	0.57187	0.67890	**0.72508**
	Stdev	0.07287	0.09071	0.06717	0.07459
Seeds	Avg	0.77048	0.76095	0.76762	**0.82810**
	Stdev	0.09162	0.11586	0.08056	0.10025
Australian	Avg	0.67087	0.63913	0.63971	**0.72928**
	Stdev	0.12386	0.12099	0.09408	0.14198
Glass	Avg	0.45047	0.44206	**0.48972**	0.47804
	Stdev	0.03201	0.04295	0.03763	0.02136
Habitit	Avg	**0.80000**	0.79871	0.79935	0.79355
	Stdev	0.01490	0.01045	0.01234	0.00000
Heart	Avg	0.68815	0.70148	0.70519	**0.78222**
	Stdev	0.10174	0.10612	0.08145	0.05627

the competitive results of all methods on Habitit dataet, we see that the GA and DE can slightly outperform MVO and PSO on Glass and Habitit datasets, respectively. This indicates that higher percentage of data points are clustered correctly in MVO-based clustering technique compared to other peers.

To recapitulate, these results verify the efficacy of the MVO compared to well-regarded optimizers such as DE, PSO, and GA in dealing with used datasets. The main reason for observed superiorities is that the MVO can make a more stable balance between diversification and intensification inclinations. Additionally, abrupt variations in the position vector and dynamic nature of *TDR* and *WEP* parameters also assist MVO in resolving possible local optima (LO) stagnation problems. Hence, it can show a successful performance in LO escaping behaviors and immature convergence drawbacks observed in DE, PSO and GA.

6   Conclusions and Future Directions

In this chapter, theoretical foundation, operations, and main strengths behind the MVO algorithm were discussed and explained. Then, a review of the several variants of the MVO algorithm were summarized. Furthermore, the main applications of the MVO algorithm in engineering, data mining, and machine learning were also thoroughly described. In addition, MVO optimizer was reformulated to deal with clustering applications. Eight clustering datasets were used to assess the efficiency of the MVO-based clustering in comparison with well regarded algorithms. The comprehensive results and analysis disclose the superiority of the MVO in terms of the optimality of results and convergence behaviors in dealing with clustering datasets.

The future studies can enhance the original exploration and exploitation operators of this method. They also can utilize the proposed MVO clustering for tackling other applications such as image segmentations, spatial databases, and spatio-temporal clustering applications.

References

1. Abusnaina, A. A., Ahmed, S., Jarrar, R., & Mafarja, M. (2018). *Training neural networks using salp swarm algorithm for pattern classification, 2*.
2. Al-Madi, N., Aljarah, I., & Ludwig, S. A. (2014). Parallel glowworm swarm optimization clustering algorithm based on mapreduce. In *2014 IEEE Symposium on Swarm Intelligence (SIS)* (pp. 1–8). IEEE.
3. Ali, E., El-Hameed, M., El-Fergany, A., & El-Arini, M. (2016). Parameter extraction of photovoltaic generating units using multi-verse optimizer. *Sustainable Energy Technologies and Assessments, 17*, 68–76.
4. Aljarah, I., AlaM, A. Z., Faris, H., Hassonah, M. A., Mirjalili, S., & Saadeh, H. (2018). Simultaneous feature selection and support vector machine optimization using the grasshopper optimization algorithm. *Cognitive Computation* (pp. 1–18).
5. Aljarah, I., Faris, H., & Mirjalili, S. (2018). Optimizing connection weights in neural networks using the whale optimization algorithm. *Soft Computing, 22*(1), 1–15.
6. Aljarah, I., Faris, H., Mirjalili, S., & Al-Madi, N. (2018). Training radial basis function networks using biogeography-based optimizer. *Neural Computing and Applications, 29*(7), 529–553.
7. Aljarah, I., & Ludwig, S. A. (2012). Parallel particle swarm optimization clustering algorithm based on mapreduce methodology. In *2012 Fourth World Congress on Nature and Biologically Inspired Computing (NaBIC)* (pp. 104–111). IEEE.
8. Aljarah, I., & Ludwig, S. A. (2013). Mapreduce intrusion detection system based on a particle swarm optimization clustering algorithm. In *Evolutionary Computation (CEC), 2013 IEEE Congress on* (pp. 955–962). IEEE.
9. Aljarah, I., & Ludwig, S. A. (2013). A new clustering approach based on glowworm swarm optimization. In *2013 IEEE Congress on Evolutionary Computation (CEC)* (pp. 2642–2649). IEEE.
10. Aljarah, I., & Ludwig, S. A. (2013). Towards a scalable intrusion detection system based on parallel PSO clustering using mapreduce. In *Proceedings of the 15th Annual Conference Companion on Genetic and Evolutionary Computation* (pp. 169–170). ACM.
11. Aljarah, I., & Ludwig, S. A. (2016). A scalable mapreduce-enabled glowworm swarm optimization approach for high dimensional multimodal functions. *International Journal of Swarm Intelligence Research (IJSIR), 7*(1), 32–54.
12. Aljarah, I., Mafarja, M., Heidari, A. A., Faris, H., Zhang, Y., & Mirjalili, S. (2018). Asynchronous accelerating multi-leader salp chains for feature selection. *Applied Soft Computing, 71*, 964–979.
13. Aminisharifabad, M., Yang, Q., & Wu, X. (2018). A penalized Autologistic regression with application for modeling the microstructure of dual-phase high strength steel. *Journal of Quality Technology*, in-press.

14. Barham, R., & Aljarah, I. (2017). Link prediction based on whale optimization algorithm. In *2017 International Conference on New Trends in Computing Sciences (ICTCS)* (pp. 55–60). IEEE.
15. Benmessahel, I., Xie, K., & Chellal, M. (2017). A new evolutionary neural networks based on intrusion detection systems using multiverse optimization. *Applied Intelligence*, 1–13.
16. Boley, D., Gini, M., Gross, R., Han, E. H. S., Hastings, K., Karypis, G., et al. (1999). Partitioning-based clustering for web document categorization. *Decision Support Systems, 27*(3), 329–341.
17. Celebi, M. E. (2014). *Partitional clustering algorithms*. Springer.
18. Chitsaz, H., & Aminisharifabad, M. (2015). Exact learning of rna energy parameters from structure. *Journal of Computational Biology, 22*(6), 463–473.
19. Das, S., Abraham, A., & Konar, A. (2008). Automatic clustering using an improved differential evolution algorithm. *IEEE Transactions on Systems, Man, and Cybernetics-Part A: Systems and Humans, 38*(1), 218–237.
20. Elfattah, M. A., Hassanien, A. E., Abuelenin, S., & Bhattacharyya, S. (2019). Multi-verse optimization clustering algorithm for binarization of handwritten documents. In *Recent Trends in Signal and Image Processing* (pp. 165–175). Springer (2019).
21. Ewees, A. A., El Aziz, M. A., & Hassanien, A. E. (2017). Chaotic multi-verse optimizer-based feature selection. *Neural Computing and Applications*, 1–16.
22. Faris, H., Aljarah, I., Al-Betar, M. A., & Mirjalili, S. Grey wolf optimizer: A review of recent variants and applications. *Neural Computing and Applications*, 1–23.
23. Faris, H., Aljarah, I., Al-Madi, N., & Mirjalili, S. (2016). Optimizing the learning process of feedforward neural networks using lightning search algorithm. *International Journal on Artificial Intelligence Tools, 25*(06), 1650033.
24. Faris, H., Aljarah, I., & Al-Shboul, B. (2016). A hybrid approach based on particle swarm optimization and random forests for e-mail spam filtering. In *International Conference on Computational Collective Intelligence* (pp. 498–508). Springer.
25. Faris, H., Aljarah, I., & Mirjalili, S. (2016). Training feedforward neural networks using multi-verse optimizer for binary classification problems. *Applied Intelligence, 45*(2), 322–332.
26. Faris, H., Aljarah, I., & Mirjalili, S. (2017). Evolving radial basis function networks using moth–flame optimizer. In *Handbook of Neural Computation* (pp. 537–550). Elsevier.
27. Faris, H., Aljarah, I., & Mirjalili, S. (2018). Improved monarch butterfly optimization for unconstrained global search and neural network training. *Applied Intelligence, 48*(2), 445–464.
28. Faris, H., Aljarah, I., Mirjalili, S., Castillo, P. A., & Merelo, J. J. (2016). Evolopy: An open-source nature-inspired optimization framework in python. In *IJCCI (ECTA)* (pp. 171–177).
29. Faris, H., & Aljarah, I., et al. (2015). Optimizing feedforward neural networks using krill herd algorithm for e-mail spam detection. In *2015 IEEE Jordan Conference on Applied Electrical Engineering and Computing Technologies (AEECT)* (pp. 1–5). IEEE.
30. Faris, H., Ala'M, A. Z., Heidari, A. A., Aljarah, I., Mafarja, M., Hassonah, M. A., & Fujita, H. (2019). An intelligent system for spam detection and identification of the most relevant features based on evolutionary random weight networks. *Information Fusion, 48*, 67–83.
31. Faris, H., Hassonah, M. A., AlaM, A. Z., Mirjalili, S., & Aljarah, I. (2017). A multi-verse optimizer approach for feature selection and optimizing SVM parameters based on a robust system architecture. *Neural Computing and Applications* (pp. 1–15).
32. Faris, H., Mafarja, M. M., Heidari, A. A., Aljarah, I., AlaM, A. Z., Mirjalili, S., et al. (2018). An efficient binary salp swarm algorithm with crossover scheme for feature selection problems. *Knowledge-Based Systems, 154*, 43–67.
33. Fathy, A., & Rezk, H. (2018). Multi-verse optimizer for identifying the optimal parameters of pemfc model. *Energy, 143*, 634–644.
34. Ghatasheh, N., Faris, H., Aljarah, I., & Al-Sayyed, R. M. (2015). Optimizing software effort estimation models using firefly algorithm. *Journal of Software Engineering and Applications, 8*(03), 133.

35. Guan, Y., Ghorbani, A. A., & Belacel, N. (2003). Y-means: A clustering method for intrusion detection. In *Canadian Conference on Electrical and Computer Engineering, IEEE CCECE 2003*. vol. 2, (pp. 1083–1086). IEEE.
36. Guha, D., Roy, P. K., & Banerjee, S. (2017). Multi-verse optimisation: a novel method for solution of load frequency control problem in power system. *IET Generation, Transmission & Distribution, 11*(14), 3601–3611.
37. Hassanin, M. F., Shoeb, A. M., & Hassanien, A. E. (2017). Designing multilayer feedforward neural networks using multi-verse optimizer. In *Handbook of Research on Machine Learning Innovations and Trends* (pp. 1076–1093). IGI Global.
38. Heidari, A. A., Faris, H., Aljarah, I., & Mirjalili, S. (2018). An efficient hybrid multilayer perceptron neural network with grasshopper optimization. *Soft Computing*, 1–18.
39. Heidari, A. A., & Abbaspour, R. A. (2018). Enhanced chaotic grey wolf optimizer for real-world optimization problems: A comparative study. In *Handbook of Research on Emergent Applications of Optimization Algorithms* (pp. 693–727). IGI Global.
40. Heidari, A. A., Abbaspour, R. A., & Jordehi, A. R. (2017). An efficient chaotic water cycle algorithm for optimization tasks. *Neural Computing and Applications, 28*(1), 57–85.
41. Heidari, A. A., Abbaspour, R. A., & Jordehi, A. R. (2017). Gaussian bare-bones water cycle algorithm for optimal reactive power dispatch in electrical power systems. *Applied Soft Computing, 57*, 657–671.
42. Heidari, A. A., & Delavar, M. R. (2016). A modified genetic algorithm for finding fuzzy shortest paths in uncertain networks. In *ISPRS - International Archives of the Photogrammetry, Remote Sensing and Spatial Information Sciences XLI-B2* (299–304).
43. Heidari, A. A., & Pahlavani, P. (2017). An efficient modified grey wolf optimizer with lévy flight for optimization tasks. *Applied Soft Computing, 60*, 115–134.
44. Hu, C., Li, Z., Zhou, T., Zhu, A., & Xu, C. (2016). A multi-verse optimizer with levy flights for numerical optimization and its application in test scheduling for network-on-chip. *PloS One, 11*(12), e0167341.
45. Jain, A. K., Murty, M. N., & Flynn, P. J. (1999). Data clustering: A review. *ACM Computing Surveys (CSUR), 31*(3), 264–323.
46. Karthikeyan, K., & Dhal, P. (2017). Multi verse optimization (mvo) technique based voltage stability analysis through continuation power flow in ieee 57 bus. *Energy Procedia, 117*, 583–591.
47. Kouba, N. E. Y., Menaa, M., Hasni, M., & Boudour, M. (2018). Application of multi-verse optimiser-based fuzzy-pid controller to improve power system frequency regulation in presence of hvdc link. *International Journal of Intelligent Engineering Informatics, 6*(1–2), 182–203.
48. Kouba, N. E. Y., Menaa, M., Hasni, M., Tehrani, K., & Boudour, M. (2016). A novel optimized fuzzy-pid controller in two-area power system with hvdc link connection. In *2016 International Conference on Control, Decision and Information Technologies (CoDIT)* (pp. 204–209). IEEE.
49. Kumar, A., & Suhag, S. (2017). Effect of tcps, smes, and dfig on load frequency control of a multi-area multi-source power system using multi-verse optimized fuzzy-pid controller with derivative filter. *Journal of Vibration and Control, 1077546317724968*.
50. Kumar, A., & Suhag, S. (2017). Multiverse optimized fuzzy-pid controller with a derivative filter for load frequency control of multisource hydrothermal power system. *Turkish Journal of Electrical Engineering & Computer Sciences, 25*(5), 4187–4199.
51. Kwedlo, W. (2011). A clustering method combining differential evolution with the k-means algorithm. *Pattern Recognition Letters, 32*(12), 1613–1621.
52. Lichman, M. (2013). UCI machine learning repository. http://archive.ics.uci.edu/ml
53. Mafarja, M., Aljarah, I., Heidari, A. A., Hammouri, A. I., Faris, H., & AlaM, A. Z., et al. (2017). Evolutionary population dynamics and grasshopper optimization approaches for feature selection problems. *Knowledge-Based Systems*.
54. Mafarja, M., Aljarah, I., Heidari, A. A., Faris, H., Fournier-Viger, P., Li, X., & Mirjalili, S. (2018). Binary dragonfly optimization for feature selection using time-varying transfer functions. *Knowledge-Based Systems, 161*, 185–204.

55. Mafarja, M., & Mirjalili, S. (2017). Whale optimization approaches for wrapper feature selection. *Applied Soft Computing, 62*, 441–453.
56. Mafarja, M. M., & Mirjalili, S. (2017). Hybrid whale optimization algorithm with simulated annealing for feature selection. *Neurocomputing, 260*, 302–312.
57. Majdi, M., Abdullah, S., & Jaddi, N. S. (2015). Fuzzy population-based meta-heuristic approaches for attribute reduction in rough set theory. *World Academy of Science, Engineering and Technology, International Journal of Computer, Electrical, Automation, Control and Information Engineering, 9*(12), 2462–2470.
58. Van der Merwe, D., & Engelbrecht, A. P. (2003). Data clustering using particle swarm optimization. In *Evolutionary Computation, 2003. CEC'03 The 2003 Congress on*. vol. 1 (pp. 215–220). IEEE.
59. Meshkat, M., & Parhizgar, M. (2017). Stud multi-verse algorithm. In *Swarm Intelligence and Evolutionary Computation (CSIEC), 2017 2nd Conference on* (pp. 42–47). IEEE.
60. Mirjalili, S., Jangir, P., Mirjalili, S. Z., Saremi, S., & Trivedi, I. N. (2017). Optimization of problems with multiple objectives using the multi-verse optimization algorithm. *Knowledge-Based Systems, 134*, 50–71.
61. Mirjalili, S., Mirjalili, S. M., & Hatamlou, A. (2016). Multi-verse optimizer: A nature-inspired algorithm for global optimization. *Neural Computing and Applications, 27*(2), 495–513.
62. Ng, H., Ong, S., Foong, K., Goh, P., & Nowinski, W. (2006). Medical image segmentation using k-means clustering and improved watershed algorithm. In *Image Analysis and Interpretation, 2006 IEEE Southwest Symposium on* (pp. 61–65). IEEE.
63. Pan, W., Zhou, Y., & Li, Z. (2017). An exponential function inflation size of multi-verse optimisation algorithm for global optimisation. *International Journal of Computing Science and Mathematics, 8*(2), 115–128.
64. Rokach, L., & Maimon, O. (2005). *Clustering methods. In: Data mining and knowledge discovery handbook* (pp. 321–352). Springer.
65. Rosenberg, A., & Hirschberg, J. (2007). V-measure: A conditional entropy-based external cluster evaluation measure. *EMNLP-CoNLL, 7*, 410–420.
66. Sayed, G. I., Darwish, A., & Hassanien, A. E. (2017). Quantum multiverse optimization algorithm for optimization problems. *Neural Computing and Applications, 1–18*.
67. Sayed, G. I., Darwish, A., & Hassanien, A. E. (2018). A new chaotic multi-verse optimization algorithm for solving engineering optimization problems. *Journal of Experimental & Theoretical Artificial Intelligence, 30*(2), 293–317.
68. Shelokar, P., Jayaraman, V. K., & Kulkarni, B. D. (2004). An ant colony approach for clustering. *Analytica Chimica Acta, 509*(2), 187–195.
69. Shukri, S., Faris, H., Aljarah, I., Mirjalili, S., & Abraham, A. (2018). Evolutionary static and dynamic clustering algorithms based on multi-verse optimizer. *Engineering Applications of Artificial Intelligence, 72*, 54–66.
70. Strehl, A., Ghosh, J., & Mooney, R. (2000). Impact of similarity measures on web-page clustering. In *Workshop on Artificial Intelligence for Web Search (AAAI 2000)*. vol. 58 (p. 64).
71. Sulaiman, M. H., Mohamed, M. R., Mustaffa, Z., & Aliman, O. (2016). An application of multi-verse optimizer for optimal reactive power dispatch problems. *International Journal of Simulation-Systems, Science & Technology, 17*, 41.
72. Trivedi, I. N., Jangir, P., Jangir, N., Parmar, S. A., Bhoye, M., & Kumar, A. (2016). Voltage stability enhancement and voltage deviation minimization using multi-verse optimizer algorithm. In *Circuit, Power and Computing Technologies (ICCPCT), 2016 International Conference on* (pp. 1–5). IEEE.
73. Valenzuela, M., Peña, A., Lopez, L., & Pinto, H. (2017). A binary multi-verse optimizer algorithm applied to the set covering problem. In *Systems and Informatics (ICSAI), 2017 4th International Conference on* (pp. 513–518). IEEE.
74. Vivek, K., Deepak, M., Mohit, J., Asha, R., & Vijander, S., et al. (2018). Development of multi-verse optimizer (mvo) for labview. In *Intelligent Communication, Control and Devices* (pp. 731–739). Springer.

75. Wang, X., Luo, D., Liu, J., Wang, W., & Jie, G. (2017). Prediction of natural gas consumption in different regions of china using a hybrid mvo-nngbm model. *Mathematical Problems in Engineering, 2017.*
76. Wang, X., Luo, D., Zhao, X., & Sun, Z. (2018). Estimates of energy consumption in china using a self-adaptive multi-verse optimizer-based support vector machine with rolling cross-validation. *Energy, 152,* 539–548.
77. Zhao, H., Han, X., & Guo, S. (2016). Dgm (1, 1) model optimized by mvo (multi-verse optimizer) for annual peak load forecasting. *Neural Computing and Applications, 1–15.*

Moth-Flame Optimization Algorithm: Theory, Literature Review, and Application in Optimal Nonlinear Feedback Control Design

Seyed Hamed Hashemi Mehne and Seyedali Mirjalili

Abstract A direct numerical method for optimal feedback control design of general nonlinear systems is presented in this chapter. The problem is generally infinite dimensional. In order to convert it to a finite dimensional optimization problem, a collocation type method is proposed. The collocation approach is based on approximating the control input function as a series of given base functions with unknown coefficients. Then, the optimal control problem is converted to the problem of finding a finite set of coefficients. To solve the resulting optimization problem, a new nature-inspired optimization paradigm known as Moth Flame Optimizer (MFO) is used. Validation and evaluating of accuracy of the method are performed via implementing it on some well known benchmark problems. Investigations presented in this chapter reveals the efficiency of the method and its benefits with respect to other numerical approaches. The chapter also consideres an in-depth lierratur review and analysis of MFO.

1 Introduction

One of the fundamental and practical problems in control theory is to design controllers that minimize a predefined performance index estimating the success of the control inputs. This problem is known as optimal control problem. Exact or analytical solution may be found using necessary optimality conditions for linear or other simple systems. However, solving the equivalent boundary value problems arising

S. H. H. Mehne
Aerospace Research Institute, 1465774111 Tehran, Iran
e-mail: hmehne@ari.ac.ir

S. Mirjalili (✉)
School of Information and Communication Technology, Griffith University,
Nathan Campus, Brisbane, QLD 4111, Australia
e-mail: seyedali.mirjalili@griffithuni.edu.au

© Springer Nature Switzerland AG 2020
S. Mirjalili et al. (eds.), *Nature-Inspired Optimizers*, Studies in Computational
Intelligence 811, https://doi.org/10.1007/978-3-030-12127-3_9

from optimality conditions is a formidable task in general, especially for non-linear systems. Therefore, approximate or numerical methods have been developed for solving optimal control problems by researchers. Due to the dynamic of the system and available tools, different methods exit which may be categorized in three classes as follow:

Approximating the Dynamical System: One of the popular procedures in finding the solution of optimal control problems is approximating or converting the underlying dynamical system with one or a set of simple dynamical systems with known or easy to find solutions. As example, a method for approximating nonlinear optimal feedback controllers for a class of general nonlinear systems was proposed in [1]. The method is based on successive linear time varying approximations. In [2], the underlying system was transformed into the Jurdjevic-Quinn type system using singular perturbation and zero dynamics theory. Then, with characteristic of this type of systems and Lyapunov theory, an optimal stabilization control law was derived. A method for approximate the stabilizing control law as the solution of another system of two partial differential equations was also presented in [3].

Indirect Methods: In the case that sufficient conditions of optimality such as Pontryagins maximum principle or Hamilton-Jacobi-Bellman equations are utilized, the method is called indirect. Using these conditions transfers the optimal control to a two point boundary value problem which may be easier to solve. For example of indirect methods, one may address [4], where a computational method for obtaining feedback optimal control for nonlinear systems based on the Hamilton-Jacobi-Bellman equation was reported. Using Pontryagins maximum principle in [5] a method for an optimal control problem of Mars landing in powered descent was derived. The method is based on reduction of the original problem to a two point boundary problem and real-time sampling optimal feedback control theory.

Indirect methods are usually computationally inexpensive and accurate. However, the available methods for solving the boundary value problem are dependent on initial guess of the solution that restricts the domain of convergence.

Direct Methods: With direct method we means an approach that transfers the optimal control problem to an optimization problem. It may be performed by collocation (or parametrization) and discretization (or time-marching). Some recent examples of direct methods are reviewed here. A direct method for designing feedback controllers that maximize the set of points that reach a given target set in specified finite time was given in [6]. The method is based on convex optimization in the form of finite dimensional linear programming problem. In [7] a numerical method for missile formation problem is introduced that is based on direct discretization of feedback control and using nonlinear programming. A numerical method using piece-wise control action was also reported in [8] for optimal control problems. Discretization of the control function leads to a nonlinear programming problem which has been solved by gradient based methods. A pseudo-spectral method was given in [9] where, state parametrization using Lagrange polynomials at special points converts the problem to an optimization one. Ease of implementation and wider convergence range are benefits of direct methods with respect to indirect ones.

Contribution of the Present Work: In this chapter, a numerical method for solving optimal control problems will be given. The control system is closed loop and optimal feedback controller for non-linear systems is under investigation. The method is in the category of direct methods based on collocation of the control. The first step is transferring the original problem to a finite dimensional optimization problem. This is performed by looking at the controller as a combination of same known functions and unknown coefficients. The resulting problem is a finite dimensional optimization problem. The second step is to solve this optimization problem by moth flame optimization algorithm that is an efficient nature-inspired optimizer. In comparison with discretization based methods like [10], the proposed method has lower dimension. Therefore, it is fast, accurate and with simple implementation. The method is tested on some benchmark problems in order to evaluation and validation. Results show that the proposed method is effective and accurate in practice.

Organization: The rest of this chapter is organized as follows: In Sect. 2, the MFO algorithm is reviewed and explained. Sect. 3 presents a litrature review of MFO. The optimal control problem is introduced and its application are noticed in Sect. 4 and MFO is tailored for optimal control problems with feedback control. Numerical examples and results are discussed in Sect. 5. Finally some conclusion remarks and future work suggestions are given in Sect. 6.

2 Moth Flame Optimization Method

2.1 Background and Motivation

The problem of finding the best situation among different choices is formulated as an optimization problem. Such a problem has objective function, decision variables and constraints as its pillars. Traditional methods for solving optimization problems such as Newton-Raphson, quasi-Newton, and conjugated gradient are based on gradient calculation. Theses methods are usually time consuming for complex problems in real world applications due to presence of partial derivatives. Therefore, search methods like golden search, Nelder-Mead and Tabu search have been developed which are just based on the objective function evaluations.

Among search methods, meta-heuristic optimization algorithms were developed imitating physical phenomena or nature behaviors. Meta-heuristic optimizers are popular in engineering applications because they are easy to implement, do not require gradient calculation, converge to global optima, and can be utilized in a wide range of problems covering different disciplines[11]. Some of the recent developed meta-heuristics are Whale Optimization [11], Dragonfly algorithm [12], Salp Swarm Algorithm [13], Colliding Bodies Optimization [14], and Thermal exchange optimization [15].

Moth-flame optimization algorithm is also a novel nature-inspired algorithm. This algorithm simulates the navigating mechanism of moths in nature known as transverse orientation. Since its development, it has been used to solve different engineering optimization problems such as optimization of electrical power generation systems [16], estimation of model parameters in solar cells [17], estimating the weights and parameters in neural networks [18], image processing [19], forecasting on annual electricity consumption [20], optimizing the manufacturing processes [21], and bio-informatics [22, 23].

2.2 Explanation of MFO

The moth flame optimization method is introduced and explained in this section. For more details about this optimizer the reader is referred to [24].

2.3 Inspiration from Moths

As stated before, MFO is a nature-inspired optimization method proposed by S. Mirjalili in 2015. The method simulates the behavior of moths for navigating in night. They fly at night using the moon based on a mechanism called transverse orientation. Actually, the moths maintain a fixed angle to the moon when flying as shown in Fig. 1. Because of long distance from earth to moon (384,400 km), this mechanism helps them to fly in straight line. However, in the face of artificial lights such as electrical lamps or candles, using transverse orientation causes a spiral path to the light source. This is because of the short distance from moth to this sources of lights when maintaining the same angle. Being the base of the MFO method, this fact was depicted in Fig. 2. In the case of planar flying for one dimensional case, the mathematical equation of such a logarithmic spiral is as follows:

$$S = De^{bt}cos(2\pi t) + F \tag{1}$$

where, F is the position of flame, M is the position of moth, $D = |M - F|$ is the distance between moth and flame, and $t \in [-1, 1]$ is a time variable. The above equation constructs the path from moth to flame when t decreases from -1.5 to 1 as shown in Fig. 3. The constant b determines the rate of reaching to flame.

2.4 The MFO Algorithm

Let assume that the underlying optimization problem is d-dimensional dimensional, that is there exit d decision variables with desired optimum values. As the MFO is a population-base method, we need some search agents. A set of-n moths is the population of search agents, where each moth keeps a value of a d dimensional

Fig. 1 Transverse orientation of moths

Fig. 2 Spiral path of flying around a close light source

vector as a candidate solution for the problem. In other words, moths are flying in the d dimensional space and their positions are possible solutions. Therefore, the set of-moths and positions may be described by the following $n \times d$ matrix:

Fig. 3 A typical logarithmic
spiral path around a flame
with some positions with
respect to *t*

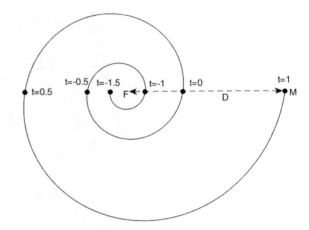

$$M = \begin{bmatrix} M_{11} & M_{12} & \cdots & \cdots & M_{1d} \\ M_{21} & M_{22} & \cdots & \cdots & M_{2d} \\ \vdots & \vdots & \vdots & \vdots & \vdots \\ M_{n1} & M_{n2} & \cdots & \cdots & M_{nd} \end{bmatrix} \qquad (2)$$

All fitness values are also stored in a vector as:

$$OM = \begin{bmatrix} OM_1 \\ OM_2 \\ \vdots \\ OM_n \end{bmatrix} \qquad (3)$$

where, each element shows the fitness value of the corresponding moth's position.
There are also a set of flames indicating the best solution that every moths has found
so far. Flames have a similar matrix representation as follows:

$$F = \begin{bmatrix} F_{11} & F_{12} & \cdots & \cdots & F_{1d} \\ F_{21} & F_{22} & \cdots & \cdots & F_{2d} \\ \vdots & \vdots & \vdots & \vdots & \vdots \\ F_{n1} & F_{n2} & \cdots & \cdots & F_{nd} \end{bmatrix} \qquad (4)$$

In the same way, the set of the corresponding best fitness values is stored to a vector
as follows:

$$OF = \begin{bmatrix} OF_1 \\ OF_2 \\ \vdots \\ OF_n \end{bmatrix} \qquad (5)$$

Therefore, flames can be interpreted as flags or pins in the search space that are dropped by moths when searching. A nominal moth flies and searches around a flame for better solution. It updates the flam position when such a better solution is found.

From mathematical viewpoint, the MFO consists of three parts as:

$$MFO = (I, P, T) \qquad (6)$$

where, $I : \emptyset \rightarrow \{M, OM\}$ is a function that generates a random population of moths and corresponding fitness values at the beginning. $P : M \rightarrow M$ is the update function that receives the current M matrix and updates it each iteration. $T : M \rightarrow \{True, False\}$ is a termination function that returns $True$ when the termination condition happens and returns $False$ when termination doesn't attain.

Therefore, the MFO is expressed in brief as follows:

$M = I()$; **while** $T(M)$ *is equal to False*
   $M = P(M)$;
end

The function I generates random initial solutions and calculates the objective function values. The following method is utilized as default:

for $i = 1 : n$
  for $j = 1 : d$
   $M_{ij} = (ub_i - lb_i) * rand() + lb(i)$
  end
end

where, ub and lb are two vectors including upper and lower bound for decision variables as follow:

$$ub = [ub_1, ub_2, \ldots, ub_d] \qquad (7)$$
$$lb = [lb_1, lb_2, \ldots, lb_d] \qquad (8)$$

After function I generates the initial solution, the iteration can be started and the function P is run in iterations until T function announces the termination. The P function is the main part of the MFO that moves moths around the search space and updates the position of flames. The position of i-th moth with respect to j-th flame is updates based on the following equations:

$$M_i = S(M_i, F_j) = D_{ij}e^{bt}\cos(2\pi t) + F_j \qquad (9)$$
$$D_{ij} = |M_i - F_j| \qquad (10)$$

As stated before, this is a logarithmic spiral from moth to the flame. Figure 4 shows four different possible points in a typical spiral in one dimension (dashed black lines). These positions can be chosen in the next iteration as moth position (horizontal line) around the flame (vertical line) depending the random variables. Therefore, a moth can explore and exploit the search space. Exploration in general encourages the candidate solutions to change its moving direction. Consequently, the exploration helps methods to prevent from trapping in local optima. Here, the exploration occurs when the new position is outside the space between the moth and flam as can be seen in the arrows labeled by 1, 3, and 4. The exploitation phase tends to improve the quality of solutions by searching locally around the obtained promising solutions in the exploration milestone. Exploitation in MFO happens when the new position lies inside the space between the moth and flame as can be observed in the arrow labeled by 2. It is to be noted that each moth can update its position with respect to only one flame at each iteration. After updating the list of flames, they are sorted based on their fitness values from the best to the worst. Then the moths update their positions according to corresponding flame. This means that for example the first moth is updated with respect to the flame with best fitness value, and so on. More over, to prevent form degrading the exploitation of the best promising solution, an adapting mechanism was proposed in [24]. This mechanism decreases the number of flames during iteration based on the following relation:

$$\text{Flame number} = round\left(N - l * \frac{N-1}{T}\right) \tag{11}$$

where, l is the iteration counter, T is the maximum number of iterations, and N is the maximum number (or initial number) of flames. Now, by the above mentioned description, the algorithm of function P is as follows (Fig. 5):

Update flame number using Eq. (9.11)
$OM = FitnessFunction(M)$;
if *iteration* $== 1$
  $F = sort(M)$;
  $OF = sort(OM)$;
else
  $F = sort(M_{t-1}, M_t)$;
  $OF = sort(M_{t-1}, M_t)$;
end
for $i = 1 : n$
  for $j = 1 : d$
    Update r and t
    Calculate D_{ij} *using Eq.*(9.10) *with respect to the corresponding moth*
    Update M_{ij} *using Eq.*(9.9) *with respectto the corresponding moth*
  end
end

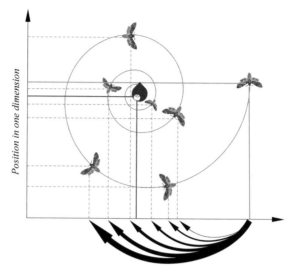

Fig. 4 Some of the possible positions that can be reached by a moth with respect to a flame using the logarithmic spiral [24]

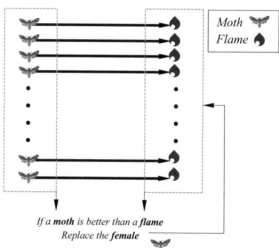

Fig. 5 Moth-Flame assignment [24]

3 Literature Review of MFO

This section provides the literature review of the MFO algorithm. It first covers different variants of this algorithm to solve problems of different types. Then, the applications of this algorithm in diverse fields are given.

3.1 Variants

This original version of MFO has been designed to solve problems with continuous variables. To solve binary problems with this algorithm, Reddy et al. [25] employed a sigmoid function to map continuous domain to binary domain. Both continuous and binary MFO can solve single-objective problems. This motivated the attempt of several researchers to developed the multi-objective version of this algorithm. For instance, a non-dominated sorting multi-objective MFO is proposed in [26]. In a similar work, an archive is used to require MFO to solve multi-objective problems [27]. Another variant of MFO was proposed in [28, 29], in which several penalty functions were used to solve constrained problems using MFO.

There are also several works in the literature to improve exploration and/or exploitation of MFO. In [30], an opposition-based learning version of MFO was propsoed by considering the opposite location of moths during the optimization process. There are three works that use chaotic maps to tune different controlling paramters of MFO to improve its performance as well [31–33]. Several Levy flights are also used to mostly boost exploration and local optima avoidance of the MFO algorithm [34, 35].

Another direction that several authors have taken is to hybridize MFO with other algorithms. There are different ways to hybridize two algorithms in the field of evolutionary computation. Most of them are used to improve the performance of MFO as follows:

- Hybrid MFO and Simulated Annealing [36]
- Hybrid MFO and Particle Swarm Optimization [37–39]
- Hybrid MFO and Gravitational Search Algorithm [40]
- Hybrid MFO and Firefly Algorithm [41]

Another popular area that MFO has been largely is Machine Learning. The problem of training different machine learning techniques is a challenging optimization problem. Gradient-based algorithms normally fail to solve such problems since they suffer from local optima stagnation and they require gradiant information of the problem. For instance, the MFO algorithm is used to find the optimal values for the main parameters of Support Vector Machine in [42]. This algorithm has been also used to train NN including finding optimal values for the connections weights, biases, and number of hidden nodes [43–46].

3.2 Applications

MFO has been applied to a large number of applications in diverse fields. The following list shows some of them:

- Breast cancer detection [47, 48]
- Optimal sizing and placement of capacitors in radial distribution systems [49]

- AGC system design [50–53]
- Clustering for Internet of Things [54, 55]
- Automatic test generation technique for software testing [56, 57]
- Dynamic economic load dispatch problem [58, 59]
- Geographic atrophy segmentation for SD-OCT images [60]
- Power dispatch problems [61–64]
- Optimal reactive power dispatch problem[65]
- Optimal allocation of shunt capacitor banks in radial distribution systems [66]
- PID controller for optimal control of active magnetic bearing System [67]
- Devices allocation in power systems [68]
- Optimal allocation of distributed generations and capacitor banks in distribution grids [69]
- Profit maximization with integration of wind farm [70]
- Dynamic performance enhancement for wind energy conversion [71]
- Web service composition in cloud computing [72]
- Image segmentation (kidney images, satellite images, etc.) [73–76]
- Optimization of water resources [77]
- Antenna array design [78]
- Optimal power tracking [79]
- Accuracy control of contactless laser sensor system [80]
- Optimal placement and sizing of distributed generation units for power loss Reduction [81]
- Range image registration [82]
- PID controller design [83]
- Alzheimers disease diagnosis [84]
- Optimal bidding strategy under transmission congestion in deregulated power market [85]
- Production planning in petrochemical industries [86]
- Terrorism prediction [87]
- Optical network unit placement in Fiber-Wireless (FiWi) access network [88]
- Hybrid power generation systems [89]
- Arabic handwritten letter recognition [90]
- Forecasting [91]
- Feature selection [92]

4 MFO for Optimal Control Problems

In this section, the form of the underlying optimal control problem is introduced. Then the method of collocation is explained and the way of implementing the MFO algorithm is discussed.

4.1   Problem Definition

The problem of designing controllers to optimally stabilize an equilibrium of a control system is a problem in the field of control with many real world applications. When the control law is based on knowledge of the system response, the controller called feedback. Feedback control is preferred over open loop solutions due to robustness to disturbances and reduction in computational time [93]. Some applications that have been reported in literature are rotor speed control in wind turbines [94], optimal stabilization and attitude tracking of satellites [2], coordinate control of turbine boilers [95], and optimal chemotherapy administration for the cancer treatment [1]. The problem of designing such a of control law is investigated in sequel.

Let us revisit the general form of a non-linear optimal control problem in which we intend to minimize a cost functional as

$$J(u) = h(x(T_f)) + \int_0^{T_f} f(x, u)dt \tag{12}$$

where, the control input $u \in \mathbb{R}^m$ and state vector $x \in \mathbb{R}^n$ are related to each other by the following dynamical system and initial condition:

$$\dot{x} = g(t, x, u), \quad x(0) = x_0 \tag{13}$$

Some of special and practical cases of this problem are linear quadratic cost function as:

$$J(u) = \frac{1}{2}x(T_f)^T H x(T_f) + \frac{1}{2}\int_0^{T_f} (x(t)^T Q x(t) + u(t)^T R u(t))dt \in \mathbb{R} \tag{14}$$

where, $H, Q \in \mathbb{R}^{n \times n}$ are positive semidefinite matrices and $R \in \mathbb{R}^{m \times m}$ is a positive definite matrix.

Linear systems in which,
$$\dot{x} = Ax(t) + Bu(t) \tag{15}$$

where, $A \in \mathbb{R}^{n \times n}$ and $B \in \mathbb{R}^{n \times m}$ are given and may be constant in invariant systems or variable with time in time varying systems.

Affine non-linear systems that has the following form:

$$\dot{x} = G(x(t)) + g(x(t))u(t) \tag{16}$$

The control values may be bounded as:

$$u_l \le u(t) \le u_b \quad \forall t \in [0, T_f] \tag{17}$$

where, u_l and u_b are constant vectors with the same dimension as $u(t)$.

4.2   Collocation

In the case of searching for feedback optimal control, the control can be considered as a function of the state, that is $u = u(x)$. For control parametrization we need a set of independent functions as base functions. They may be monomials, Chebyshev polynomials, geometric or exponential functions. Let

$$\Phi(x) = \left[\phi_1(x) \ \phi_2(x) \ \cdots \ \phi_d(x) \right]^T \tag{18}$$

is such a base. Then, the parametrized control function is a linear combination of the elements of this base. Then, a typical control function has the following form:

$$u^d(x) = \sum_{i=1}^{d} c_i \phi_i(x) = c\Phi(x)^T \tag{19}$$

where, $c = [c_1, c_2, \ldots, c_d]$ is a real value vector of coefficients determining the control. In other words, every selection of these coefficients constructs a control function. As the control is selected, the corresponding state will be obtained by solving Eq. (13) numerically for this control input. By this transformation, the problem of looking for the best control function is reduced to the problem of looking in the following subset of d-dimensional Euclidean space:

$$C = \{c = (c_1, c_2, \ldots, c_d) \in \mathbb{R}^n \mid u_l \le u^d(x) \le u_b\} \tag{20}$$

On the other hand, the objective function will be dependent implicitly on c and therefore, the original problem is transfered to an optimization problem. In the proposed method, the MFO is utilized to solve this optimization problem. Figure 6 describes the method as a block diagram showing the position of MFO in generating the vector c of coefficient, managing the iterations and stopping condition. In the next section the method will be evaluated via practical and standard case studies.

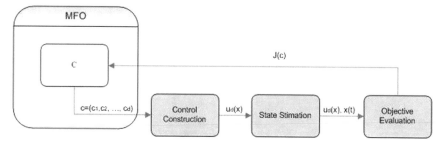

Fig. 6 Flowchart of the proposed method

5 Numerical Examples

In this section, the method of Sect. 4 is implemented on some benchmark optimal control problems to evaluate performance, accuracy and efficiency of this method.

5.1 Example 1. Rotational/Translational Actuator

Rotational/Translational actuator (RTAC) is a well known nonlinear benchmark problem in the field of feedback control. It has been used to validated and evaluate the numerical and analytical methods for deriving control laws in literature. As depicted schematically in Fig. 7, RTAC is a mechanical system consisting of a translational oscillator with an eccentric rotational proof-mass actuated by a DC motor. The rotational actuator is used to stabilize the translational motion. It was originally proposed to represent a simplified model of a dual spin spacecraft. Due to different application of this system, it has been studied as an individual system in literature. For example, in [96] a least square method was applied to estimating the parameters of RTAC system and by using experimental data, a linear system representation for RTAC was introduced using system identification. Dynamical analysis and designing of stabilizing controls of RTAC system in the case that the installation plane is not horizontal was reported in [97]. The RTAC system is also used to check the new methods in control. For example, it was used as a benchmark problem in [98] for evaluating a fuzzy reinforced learning based controller. In the present work, the proposed method is applied on this system for evaluation.

Fig. 7 Translational oscillator with rotational actuator [96]

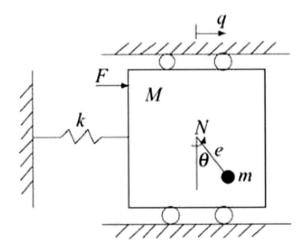

The RTAC motion in non-dimensional form is represented by the following four dimensional nonlinear system of equations [99]:

$$\dot{x}_1 = x_2 \tag{21}$$

$$\dot{x}_2 = \frac{-x_1 + \varepsilon x_4^2 \sin x_3}{1 - \varepsilon^2 \cos^2 x_3} + \frac{-\varepsilon \cos x_3}{1 - \varepsilon^2 \cos^2 x_3} u \tag{22}$$

$$\dot{x}_3 = x_4 \tag{23}$$

$$\dot{x}_4 = \frac{\varepsilon \cos x_3 (x_1 - \varepsilon x_4^2 \sin x_3)}{1 - \varepsilon^2 \cos^2 x_3} + \frac{1}{1 - \varepsilon^2 \cos^2 x_3} u \tag{24}$$

where, disturbance is ignored and
x_1 and x_2 are normalized position and velocity of the platform.
x_3 and x_4 are the angular position and velocity of the rotating mass.
u is the non-dimensional control torque.

The problem was solved by the proposed method and using MFO. The base functions for parametrization were chosen similar to [100]. They are x_1, x_2, x_3 and x_4 with pairwise combinations of degree lower than 3. The MFO parameters, including number of iterations, b, and number of moths are respectively 500, 1, and 30. Table 1 shows the base functions and the resulting coefficient for each of them in the optimal combination. The resulting feedback control function was shown in Fig. 8. As expected it has an oscillating form which damps with time. Using the control function of Fig. 8 as a control input to the system (21)–(24) and solving equations numerically by Runge-Kutta method, the system responses were found as depicted in Fig. 9. It is clear that the resulting control stabilizes all the states as desired. This shows that the proposed method works well in this case and the results are valid.

Table 1 Base functions and their resulting optimal coefficients for Example 1

i	$\phi(i)$	c_i	i	$\phi(i)$	c_i
1	x_1	5.0000	10	$x_2 x_3$	5.0000
2	x_2	−1.1159	11	$x_2 x_4$	−3.2979
3	x_3	−1.1902	12	x_3^2	4.1480
4	x_4	−4.0401	13	$x_3 x_4$	4.3387
5	x_1^2	1.0678	14	x_4^2	−0.1579
6	$x_1 x_2$	5.0000	15	x_1^3	−5.0000
7	$x_1 x_3$	−5.0000	16	x_2^3	−5.0000
8	$x_1 x_4$	−0.9885	17	x_3^3	−5.0000
9	x_2^2	−3.7699	18	x_4^3	5.0000

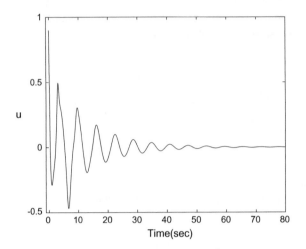

Fig. 8 The resulting optimal feedback control of the RTAC system

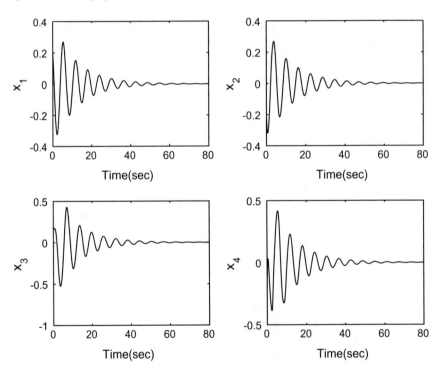

Fig. 9 The response of the RTAC system to the resulting control function

5.2 Example 2. F-8 Aircraft

The equations of F-8 dynamics in certain conditions is descried by the following control system [101]:

$$\dot{x}_1 = -0.877x_1 + x_3 - 0.088x_1x_3 + 0.47x_1^2 - 0.019x_2^2 - x_1^2x_3 + 3.846x_1^3 - 0.215u$$
$$+0.28x_1^2u + 0.47x_1u^2 + 0.63u^3 \qquad (25)$$
$$\dot{x}_2 = x_3 \qquad (26)$$
$$\dot{x}_3 = -4.208x_1 - 0.396x_3 - 0.47x_1^2 - 3.564x_1^3 - 20.967u + 6.265x_1^2u + 46x_1u^2$$
$$+61.4u^3 \qquad (27)$$

where, x_1 denotes the angle of attack (rad), x_2 is the pitch angle (rad), and x_3 shows the pitch rate (rad/s). The control input u is related to the elevator angle (rad). The aim of this example is to provide a control law that stabilizes an initial disturbance in the angle of attack in addition to minimizing LQR performance index as EQ. (14). The problem has been solved by the proposed method with data given from [101] such as $T_f = 5(sec)$, $H = 0.1I_3$, $Q = 0.01I_{3\times3}$, and $R = 1$. Initial condition indicating the disturbance is $x(0) = (0.56, 0, 0)$. For parametrization, 12 different forms of monomials were utilized. Number of iterations was 100, b were set to 1, and the number of moths was 30. Table 2 shows the base functions and resulting optimal coefficients. The control function and the response of the system of the first channel have been depicted in Fig. 10. It turns out from the figure that the control function is successful in stabilizing the angle of attack and elimination of the initial disturbances. The optimal performance index was $J^* = 0.035$ and results are agreed with the results of [101]. Therefore, the proposed method has the ability and accuracy to derive control laws in practical, complex, and nonlinear systems.

Table 2 Base functions and their resulting optimal coefficients for Example 2

i	$\phi(i)$	c_i	i	$\phi(i)$	c_i
1	x_1	0.1103	10	x_2^2	−0.2600
2	x_2	0.1441	11	x_2x_3	−0.2600
3	x_3	−0.0574	12	x_3^2	0.0800
4	x_1^2	0.2600	14	x_1^3	0.2600
5	x_1x_2	−0.1271	15	x_2^3	−0.2600
6	x_1x_3	0.2600	16	x_3^3	0.1174

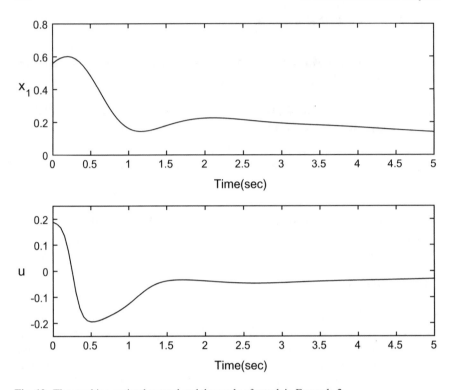

Fig. 10 The resulting optimal control and the angle of attack in Example 2

6 Conclusion

A direct numerical method for obtaining optimal feed back controls was presented. The method uses control parametrization and benefits from moth-flame optimization algorithm features. Control parametrization decreases the size of the optimization problem in comparison with discretization in the time domain. Therefore, the resulting method has lower computational complexity, it is easy to implement, fast, and doesn't care about system nonlinearities. It has been implemented and tested on two practical problems. Results show that the method could derive stabilizing controls in optimum situations. For further works, trying it on GPU structures is suggested.

Acknowledgements Authors would like to thank Mr. Farhad Karimzadeh for performing some of the graphical tasks.

References

1. Unal, C., & Salamci, M. U. (2018). Drug administration in cancer treatment via optimal nonlinear state feedback gain matrix design. *IFAC Papersonline, 50*, 9979–9984.
2. Zhang, B., Liu, K., & Xiang, J. (2013). A stabilized optimal nonlinear feedback control for satellite attitude tracking. *Aerospace Science and Technology, 27*, 17–24.
3. Mylvaganam, T., & Sassano, M. (2017). Approximate optimal control via measurement feedback for a class of nonlinear systems. *IFAC Papersonline, 50*, 15391–15396.
4. Zhu, J. (2017). A feedback optimal control by Hamilton-Jacobi-Bellman equation. *European Journal of Control, 37*, 70–74.
5. Zheng, Y., & Cui, H. (2015). Optimal nonlinear feedback guidance algorithm for Mars powered descent. *Aerospace Science and Technology, 45*, 359–366.
6. Majumdar, A., Vasudevan, R., Tobenkin, M. M., & Tedrake, R. (2014). Convex Optimization of nonlinear feedback controllers via occupation measures. *The International Journal of Robotics Research, 33*, 1209–1230.
7. Yun-jie, W., Futao, Z., & Chuang, S. (2017). Optimal discretization of feedback control in missile formation. *Aerospace Science and Technology, 67*, 456–472.
8. Armaoua, A., & Ataei, A. (2014). Piece-wise constant predictive feedback control of nonlinear systems. *Journal of Process Control, 24*, 326–335.
9. Xiao-Jun, T., Jian-Li, W., & Kai, C. (2015). A Chebyshev-Gauss pseudospectral method for solving optimal control problems. *Acta Automatica Sinica, 41*, 1778–1787.
10. Mehne, S. H. H., & Mirjalili, S. (2018). A parallel numerical method for solving optimal control problems based on whale optimization algorithm. *Knowl-Based System, 151*, 114–123.
11. Mirjalili, S., & Lewis, A. (2016). The whale optimization algorithm. *Advances in Engineering Software, 95*, 51–67.
12. Mirjalili, S. (2016). Dragonfly algorithm: A new meta-heuristic optimization technique for solving single-objective, discrete, and multi-objective problems. *Neural Computing and Applications, 27*, 1053–1073.
13. Mirjalili, S., Gandomi, A. H., Mirjalili, S. Z., Saremi, S., Faris, H., & Mirjalili, S. M. (2017). Salp Swarm Algorithm: A bio-inspired optimizer for engineering design problems. *Advances in Engineering Software, 114*, 163–191.
14. Kaveh, A., & Mahdavi, V. R. (2014). Colliding bodies optimization: A novel meta-heuristic method. *Computers Structures, 139*, 18–27.
15. Kaveh, A., & Dadras, A. (2017). A novel meta-heuristic optimization algorithm: Thermal exchange optimization. *Advances in Engineering Software, 110*, 69–84.
16. Mohamed, A. A., Mohamed, Y. S., El-Gaafary, A. A. M., & Hemeida, A. M. (2017). Optimal power flow using moth swarm algorithm. *Electric Power Systems Research, 142*, 190–206.
17. Allam, D., Yousri, D. A., & Eteiba, M. B. (2016). Parameters extraction of the three diode model for the multi-crystalline solar cell/module using Moth-flame optimization algorithm. *Energ Convers, 123*, 535–54.
18. Yamany, W., Fawzy, M., Tharwat, A., & Hassanien, A. E. (2016). Moth-flame optimization for training Multi-Layer Perceptrons. In *2015 11th International Computer Engineering Conference*. https://doi.org/10.1109/ICENCO.2015.7416360.
19. Abd El Azizab, M., Ewees, A. A., & Hassanien, A. E. (2017). Whale optimization algorithm and Moth-flame optimization for multilevel thresholding image segmentation. *Expert Systems with Applications, 83*, 242–256.
20. Zhao, H., Zhao, H., & Guo, S. (2016). Using GM (1,1) Optimized by MFO with rolling mechanism to forecast the electricity consumption of inner mongolia. *Applied Sciences, 6*. https://doi.org/10.3390/app6010020.
21. Yildiz, B. S., & Yildiz, A. R. (2017). Moth-flame optimization algorithm to determine optimal machining parameters in manufacturing processes. *Materials Testing, 59*, 425–429.
22. Chitsaz, H., & Aminisharifabad, M. (2015). Exact learning of rna energy parameters from structure. *Journal of Computational Biology, 22*(6), 463–473.

23. Aminisharifabad, M., Yang, Q. & Wu, X. (2018). A penalized Autologistic regression with application for modeling the microstructure of dual-phase high strength steel. *Journal of Quality Technology*. in-press.
24. Mirjalili, S. (2015). Moth-flame optimization algorithm: A novel nature-inspired heuristic paradigm. *Knowl-Based System, 89*, 228–249.
25. Reddy, S., Panwar, L. K., Panigrahi, B. K., & Kumar, R. (2018). Solution to unit commitment in power system operation planning using binary coded modified moth flame optimization algorithm (BMMFOA): A flame selection based computational technique. *Journal of Computational Science, 25*, 298–317. Multi-objective MFO.
26. Savsani, V., & Tawhid, M. A. (2017). Non-dominated sorting moth flame optimization (NS-MFO) for multi-objective problems. *Engineering Applications of Artificial Intelligence, 63*, 20–32.
27. Nanda, S. J. (2016, September). Multi-objective moth flame optimization. In *2016 International Conference on Advances in Computing, Communications and Informatics (ICACCI)* (pp. 2470–2476). IEEE.
28. Jangir, N., Pandya, M. H., Trivedi, I. N., Bhesdadiya, R. H., Jangir, P., & Kumar, A. (2016, March). Moth-Flame Optimization algorithm for solving real challenging constrained engineering optimization problems. In *Electrical, Electronics and Computer Science (SCEECS), 2016 IEEE Students' Conference on* (pp. 1–5). IEEE.
29. Bhesdadiya, R. H., Trivedi, I. N., Jangir, P., & Jangir, N. (2018). Moth-flame optimizer method for solving constrained engineering optimization problems. In *Advances in Computer and Computational Sciences* (pp. 61–68). Springer, Singapore.
30. Apinantanakon, W., & Sunat, K. (2017, July). OMFO: A new opposition-based moth-flame optimization algorithm for solving unconstrained optimization problems. In *International Conference on Computing and Information Technology* pp. 22–31). Springer, Cham.
31. Emary, E., & Zawbaa, H. M. (2016). Impact of chaos functions on modern swarm optimizers. *PloS One, 11*(7), e0158738.
32. Wang, M., Chen, H., Yang, B., Zhao, X., Hu, L., & Cai, Z., et al. (2017). Toward an optimal kernel extreme learning machine using a chaotic moth-flame optimization strategy with applications in medical diagnoses. *Neurocomputing, 267*, 69–84.
33. Guvenc, U., Duman, S., & Hnsloglu, Y. (2017, July). Chaotic moth swarm algorithm. In *2017 IEEE International Conference on INnovations in Intelligent SysTems and Applications (INISTA)* (pp. 90–95). IEEE.
34. Li, Z., Zhou, Y., Zhang, S., & Song, J. (2016). Lvy-flight moth-flame algorithm for function optimization and engineering design problems. *Mathematical Problems in Engineering*.
35. Trivedi, I. N., Bhesdadiya, R. H., Pandya, M. H., Jangir, N., Jangir, P., & Ladumor, D. Implementation of meta-heuristic levy flight moth-flame optimizer for solving real challenging constrained engineering optimization problems.
36. Sayed, G. I., & Hassanien, A. E. (2018). A hybrid SA-MFO algorithm for function optimization and engineering design problems. *Complex & Intelligent Systems*, 1–18.
37. Bhesdadiya, R. H., Trivedi, I. N., Jangir, P., Kumar, A., Jangir, N., & Totlani, R. (2017). A novel hybrid approach particle swarm optimizer with moth-flame optimizer algorithm. In *Advances in Computer and Computational Sciences* (pp. 569-577). Springer, Singapore.
38. Anfal, M., & Abdelhafid, H. (2017). Optimal placement of PMUs in algerian network using a hybrid particle SwarmMoth flame optimizer (PSO-MFO). *Electrotehnica, Electronica, Automatica, 65*(3).
39. Jangir, P. (2017). Optimal power flow using a hybrid particle Swarm optimizer with moth flame optimizer. *Global Journal of Research In Engineering*.
40. Sarma, A., Bhutani, A., & Goel, L. (2017, September). Hybridization of moth flame optimization and gravitational search algorithm and its application to detection of food quality. In *Intelligent Systems Conference (IntelliSys), 2017* (pp. 52–60). IEEE.
41. Zhang, L., Mistry, K., Neoh, S. C., & Lim, C. P. (2016). Intelligent facial emotion recognition using moth-firefly optimization. *Knowledge-Based Systems, 111*, 248–267.

42. Li, C., Li, S., & Liu, Y. (2016). A least squares support vector machine model optimized by moth-flame optimization algorithm for annual power load forecasting. *Applied Intelligence, 45*(4), 1166–1178.
43. Yamany, W., Fawzy, M., Tharwat, A., & Hassanien, A. E. (2015, December). Moth-flame optimization for training multi-layer perceptrons. In *Computer Engineering Conference (ICENCO), 2015 11th International* (pp. 267–272). IEEE.
44. Faris, H., Aljarah, I., & Mirjalili, S. (2017). Evolving radial basis function networks using MothFlame optimizer. In *Handbook of Neural Computation* (pp. 537–550).
45. Dosdoru, A. T., Boru, A., Gken, M., zalc, M., & Gken, T. (2018). Assessment of hybrid artificial neural networks and Metaheuristics for stock market forecasting. *ukurova niversitesi Sosyal Bilimler Enstits Dergisi, 27*(1), 63–78.
46. Kaur, N., Rattan, M., & Gill, S. S. (2018). Performance optimization of Broadwell-Y shaped transistor using artificial neural network and Moth-flame optimization technique. *Majlesi Journal of Electrical Engineering, 12*(1), 61–69.
47. Sayed, G. I., Soliman, M., & Hassanien, A. E. (2016). Bio-inspired swarm techniques for thermogram breast cancer detection. In *Medical Imaging in Clinical Applications* (pp. 487–506). Springer, Cham.
48. Sayed, G. I., & Hassanien, A. E. (2017). Moth-flame swarm optimization with neutrosophic sets for automatic mitosis detection in breast cancer histology images. *Applied Intelligence, 47*(2), 397–408.
49. Diab, A. A. Z., & Rezk, H. Optimal sizing and placement of capacitors in radial distribution systems based on Grey Wolf, Dragonfly and MothFlame optimization algorithms. *Iranian Journal of Science and Technology, Transactions of Electrical Engineering*, 1–20.
50. Mohanty, B. (2018). Performance analysis of moth flame optimization algorithm for AGC system. *International Journal of Modelling and Simulation*, 1–15.
51. Mohanty, B., Acharyulu, B. V. S., & Hota, P. K. (2018). Mothflame optimization algorithm optimized dualmode controller for multiarea hybrid sources AGC system. *Optimal Control Applications and Methods, 39*(2), 720–734.
52. Barisal, A. K., & Lal, D. K. (2018). Application of moth flame optimization algorithm for AGC of multi-area interconnected power systems. *International Journal of Energy Optimization and Engineering (IJEOE), 7*(1), 22–49.
53. Lal, D. K., Bhoi, K. K., & Barisal, A. K. (2016, October). Performance evaluation of MFO algorithm for AGC of a multi area power system. In *2016 International Conference on Signal Processing, Communication, Power and Embedded System (SCOPES)* (pp. 903–908). IEEE.
54. Reddy, M. P. K., & Babu, M. R. (2017). A hybrid cluster head selection model for internet of things. *Cluster Computing*, 1–13.
55. Yang, X., Luo, Q., Zhang, J., Wu, X., & Zhou, Y. (2017, August). Moth Swarm algorithm for clustering analysis. In *International Conference on Intelligent Computing* (pp. 503–514). Springer, Cham.
56. Metwally, A. S., Hosam, E., Hassan, M. M., & Rashad, S. M. (2016, October). WAP: A novel automatic test generation technique based on moth flame optimization. In *2016 IEEE 27th International Symposium on Software Reliability Engineering (ISSRE)* (pp. 59–64). IEEE.
57. Sharma, R., & Saha, A. (2017). Optimal test sequence generation in state based testing using moth flame optimization algorithm. *Journal of Intelligent & Fuzzy Systems*, (Preprint), 1–13.
58. Bhadoria, A., Kamboj, V. K., Sharma, M., & Bath, S. K. (2018). A solution to non-convex/convex and dynamic economic load dispatch problem using moth flame optimizer. *INAE Letters, 3*(2), 65–86.
59. Trivedi, I. N., Kumar, A., Ranpariya, A. H., & Jangir, P. (2016, April). Economic load dispatch problem with ramp rate limits and prohibited operating zones solve using Levy Flight Moth-Flame optimizer. In *2016 International Conference on Energy Efficient Technologies for Sustainability (ICEETS)* (pp. 442–447). IEEE.
60. Huang, Y., Ji, Z., Chen, Q., & Niu, S. (2017, September). Geographic atrophy segmentation for SD-OCT images by MFO algorithm and affinity diffusion. In *International Conference on Intelligent Science and Big Data Engineering* (pp. 473–484). Springer, Cham.

61. Mei, R. N. S., Sulaiman, M. H., Daniyal, H., & Mustaffa, Z. (2018). Application of Moth-flame optimizer and ant lion optimizer to solve optimal reactive power dispatch problems. *Journal of Telecommunication, Electronic and Computer Engineering (JTEC), 10*(1–2), 105–110.

62. Elsakaan, A. A., El-Sehiemy, R. A. A., Kaddah, S. S., & Elsaid, M. I. (2018). Economic power dispatch with emission constraint and valve point loading effect using moth flame optimization algorithm. In *Advanced Engineering Forum* (Vol. 28, pp. 139–149). Trans Tech Publications.

63. Trivedi, I. N., Parmar, S. A., Pandya, M. H., Jangir, P., Ladumor, D., & Bhoye, M. T. Optimal active and reactive power dispatch problem solution using Moth-Flame optimizer.

64. Anbarasan, P., & Jayabarathi, T. (2017). Optimal reactive power dispatch using Moth-flame optimization algorithm. *International Journal of Applied Engineering Research, 12*(13), 3690–3701.

65. Sulaiman, M. H., Mustaffa, Z., Aliman, O., Daniyal, H., & Mohamed, M. R. (2016). Application of moth-flame optimization algorithm for solving optimal reactive power dispatch problem.

66. Upper, N., Hemeida, A. M., & Ibrahim, A. A. (2017, December). Moth-flame algorithm and loss sensitivity factor for optimal allocation of shunt capacitor banks in radial distribution systems. In *Power Systems Conference (MEPCON), 2017 Nineteenth International Middle East* (pp. 851–856). IEEE.

67. Dhyani, A., Panda, M. K., & Jha, B. (2018). Moth-flame optimization-based fuzzy-PID controller for optimal control of active magnetic bearing system. *Iranian Journal of Science and Technology, Transactions of Electrical Engineering*, 1–13.

68. Saurav, S., Gupta, V. K., & Mishra, S. K. (2017, March). Moth-flame optimization based algorithm for FACTS devices allocation in a power system. In *2017 International Conference on Innovations in Information, Embedded and Communication Systems (ICIIECS)* (pp. 1–7). IEEE.

69. Tolba, M. A., Diab, A. A. Z., Tulsky, V. N., & Abdelaziz, A. Y. (2018). LVCI approach for optimal allocation of distributed generations and capacitor banks in distribution grids based on mothflame optimization algorithm. *Electrical Engineering*, 1–26.

70. Gope, S., Dawn, S., Goswami, A. K., & Tiwari, P. K. (2016, November). Profit maximization with integration of wind farm in contingency constraint deregulated power market using Moth flame optimization algorithm. In *Region 10 Conference (TENCON), 2016 IEEE* (pp. 1462–1466). IEEE.

71. Ebrahim, M. A., Becherif, M., & Abdelaziz, A. Y. (2018). Dynamic performance enhancement for wind energy conversion system using Moth-flame optimization based blade pitch controller. *Sustainable Energy Technologies and Assessments, 27*, 206–212.

72. GhobaeiArani, M., Rahmanian, A. A., Souri, A., & Rahmani, A. M. A mothflame optimization algorithm for web service composition in cloud computing: Simulation and verification. Software: Practice and Experience.

73. Khairuzzaman, A. K. M., & Chaudhury, S. (2017). Moth-flame optimization algorithm based multilevel thresholding for image segmentation. *International Journal of Applied Metaheuristic Computing (IJAMC), 8*(4), 58–83.

74. Said, S., Mostafa, A., Houssein, E. H., Hassanien, A. E., & Hefny, H. (2017, September). Moth-flame optimization based segmentation for MRI liver images. In *International Conference on Advanced Intelligent Systems and Informatics* (pp. 320–330). Springer, Cham.

75. Muangkote, N., Sunat, K., & Chiewchanwattana, S. (2016, July). Multilevel thresholding for satellite image segmentation with moth-flame based optimization. In *2016 13th International Joint Conference on Computer Science and Software Engineering (JCSSE)* (pp. 1–6). IEEE.

76. El Aziz, M. A., Ewees, A. A., & Hassanien, A. E. (2017). Whale optimization algorithm and Moth-flame optimization for multilevel thresholding image segmentation. *Expert Systems with Applications, 83*, 242–256.

77. Li, W. K., Wang, W. L., & Li, L. (2018). Optimization of water resources utilization by multi-objective moth-flame algorithm. *Water Resources Management*, 1–14.

78. Das, A., Mandal, D., Ghoshal, S. P., & Kar, R. (2018). Concentric circular antenna array synthesis for side lobe suppression using moth flame optimization. *AEU-International Journal of Electronics and Communications, 86*, 177–184.
79. Huang, L. N., Yang, B., Zhang, X. S., Yin, L. F., Yu, T., & Fang, Z. H. (2017). Optimal power tracking of doubly fed induction generator-based wind turbine using swarm mothflame optimizer. *Transactions of the Institute of Measurement and Control*, 0142331217712091.
80. Pathak, V. K., & Singh, A. K. (2017). Accuracy control of contactless laser sensor system using whale optimization algorithm and moth-flame optimization. *tm-Technisches Messen, 84*(11), 734–746.
81. Das, A., & Srivastava, L. Optimal placement and sizing of distributed generation units for power loss reduction using Moth-flame optimization algorithm.
82. Zou, L., Ge, B., & Chen, L. (2018). Range image registration based on hash map and moth-flame optimization. *Journal of Electronic Imaging, 27*(2), 023015.
83. Sahu, P. C., Prusty, R. C., & Panda, S. (2017, April). MFO algorithm based fuzzy-PID controller in automatic generation control of multi-area system. In *2017 International Conference on Circuit, Power and Computing Technologies (ICCPCT)* (pp. 1–6). IEEE.
84. Sayed, G. I., Hassanien, A. E., Nassef, T. M., & Pan, J. S. (2016, November). Alzheimers disease diagnosis based on Moth flame optimization. In *International Conference on Genetic and Evolutionary Computing* (pp. 298–305). Springer, Cham.
85. Gope, S., Dawn, S., Goswami, A. K., & Tiwari, P. K. (2016, November). Moth Flame optimization based optimal bidding strategy under transmission congestion in deregulated power market. In *Region 10 Conference (TENCON), 2016 IEEE* (pp. 617–621). IEEE.
86. Chauhan, S. S., & Kotecha, P. (2016, November). Single level production planning in petrochemical industries using Moth-flame optimization. In *Region 10 Conference (TENCON), 2016 IEEE* (pp. 263–266). IEEE.
87. Soliman, G. M., Khorshid, M. M., & Abou-El-Enien, T. H. (2016). Modified moth-flame optimization algorithms for terrorism prediction. *International Journal of Application or Innovation in Engineering and Management, 5*, 47–58.
88. Singh, P., & Prakash, S. (2017). Optical network unit placement in Fiber-Wireless (FiWi) access network by Moth-Flame optimization algorithm. *Optical Fiber Technology, 36*, 403–411.
89. Mekhamer, S. F., Abdelaziz, A. Y., Badr, M. A. L., & Algabalawy, M. A. (2015). Optimal multi-criteria design of hybrid power generation systems: A new contribution. *International Journal of Computer Applications, 129*(2), 13–24.
90. Ewees, A. A., Sahlol, A. T., & Amasha, M. A. (2017, May). A Bio-inspired moth-flame optimization algorithm for Arabic handwritten letter recognition. In *2017 International Conference on Control, Artificial Intelligence, Robotics & Optimization (ICCAIRO)* (pp. 154–159). IEEE.
91. Zhao, H., Zhao, H., & Guo, S. (2016). Using GM (1, 1) optimized by MFO with rolling mechanism to forecast the electricity consumption of inner mongolia. *Applied Sciences, 6*(1), 20.
92. Zawbaa, H. M., Emary, E., Parv, B., & Sharawi, M. (2016, July). Feature selection approach based on moth-flame optimization algorithm. In *2016 IEEE Congress on Evolutionary Computation (CEC)* (pp. 4612–4617). IEEE.
93. Patil, D., Mulla, A., Chakraborty, D., & Pillai, H. (2015). Computation of feedback control for time optimal state transfer using Groebner basis. *Systems & Control Letters, 79*, 1–7.
94. Jabbari Asl, H., & Yoon, J. (2016). Power capture optimization of variable-speed wind turbines using an output feedback controller. *Renewable Energy, 86*, 517–525.
95. Zhou, H., Chen, C., Lai, J., Lu, X., Deng, Q., Gao, X., et al. (2018). Affine nonlinear control for an ultra-supercritical coal fired once-through boiler-turbine unit. *Energy, 153*, 638–649.
96. Tavakoli, M., Taghirad, H. D., & Abrishamchian, M. (2005). Identification and robust H control of the rotational/translational actuator system. *International Journal of Control, Automation, 3*, 387–396.

97. Gao, B., & Ye, F. (2014). Dynamical analysis and stabilizing control of inclined rotational translational actuator systems. *Journal of Applied Mathematics,*. https://doi.org/10.1155/2014/598384.
98. Kumar, A., & Sharma, R. (2017). Fuzzy lyapunov reinforcement learning for non linear systems. *ISA Transactions, 67*, 151–159.
99. Bupp, R. T., Bernstein, D. S., & Coppola, V. T. (1998). A benchmark problem for nonlinear control design. *International Journal Robust Nonlinear Control, 8*, 307–310.
100. Luo, B, Wu, H. N., Huang, T, & Derong Liu, D. Data-based approximate policy iteration for affine nonlinear continuous-time optimal control design. *Automatica, 50*, 3281–3290.
101. Cimen, T., & Banks, S. P. (2004). Global optimal feedback control for general nonlinear systems with nonquadratic performance criteria. *Systems Control Letters, 53*, 327–346.

Particle Swarm Optimization: Theory, Literature Review, and Application in Airfoil Design

Seyedali Mirjalili, Jin Song Dong, Andrew Lewis and Ali Safa Sadiq

Abstract The Particle Swarm Optimization (PSO) is one of the most well-regarded algorithms in the literature of meta-heuristics. This algorithm mimics the navigation and foraging behaviour of birds in nature. Despite the simple mathematical model, it has been widely used in diverse fields of studies to solve optimization problems. There is a tremendous number of theoretical works on this algorithm too that has led to a large number of variants, improvements, and hybrids. This chapter covers the inspirations, mathematical equations, and the main algorithm of this technique. Its performance is tested and analyzed on a challenging real-world problem in the field of aerospace engineering.

1 Introduction

Meta-heuristics are high-level techniques using heuristics to solve a wide range of problems. As opposed to heuristics, they do not need any problem-dependent heuristic information. One of the main advantages of such techniques is the fact that they make few assumptions about the problem and attempt to consider them as a black box [1]. To search for optimal solutions, they sample the search space of the problem that is too large to be searched entirely.

S. Mirjalili (✉) · J. Song Dong · A. Lewis
Institute for Integrated and Intelligent Systems, Griffith University,
Nathan, Brisbane, QLD 4111, Australia
e-mail: seyedali.mirjalili@griffithuni.edu.au

J. Song Dong
Department of Computer Science, School of Computing, National University of Singapore,
Singapore, Singapore
e-mail: j.dong@griffith.edu.au

A. Lewis
e-mail: a.lewis@griffith.edu.au

A. S. Sadiq
School of Information Technology, Monash University, 47500 Bandar Sunway, Malaysia
e-mail: ali.safaa@monash.edu

© Springer Nature Switzerland AG 2020
S. Mirjalili et al. (eds.), *Nature-Inspired Optimizers*, Studies in Computational
Intelligence 811, https://doi.org/10.1007/978-3-030-12127-3_10

Meta-heuristics belong to the family of Soft Computing techniques, in which solutions can be partially true or inaccurate. This comes at the cost of unreliability in finding the best solution for a given problem. The extremely large size of some problems and failure of exact methods justify the need for such problem solving techniques since they find optimal or near optimal solutions in a reasonable time.

The field of meta-heuristics has seen a large number of algorithms mostly inspired from nature. Some of the most popular ones are Simulated Annealing (SA) [2] mimicking the annealing process in physics, Genetic Algorithms (GA) [3] inspired from the theory of evolution, Differential Evolution (DE) [4] incorporating the concepts of evolution in differential equations, Ant Colony Optimization (ACO) [5] inspired from the swarm intelligence of ants, and Particle Swarm Optimization (PSO) [6] that mimics navigation of birds.

Some of the algorithms belong to the family of swarm intelligence techniques. In this area, the main inspiration is the collective behaviour of insects, animals, or any other creatures that leads to fascinating global behaviours. There is no centralized control unit for decision making and the local interactions between individual/individual and individuals/environment lead to a global decision making.

The PSO algorithm is one of the most well-regarded swarm intelligence techniques that has been widely used in both science and industry [7, 8]. The main purpose of this chapter is to present the preliminaries, essential definitions, mathematical models, and algorithms of this technique.

2 Particle Swarm Optimization

The PSO algorithm has been inspired from the flocking behavior of birds in nature. In this algorithm, each particle is considered to be a solution for a given optimization problem. It is made of two vectors: position and velocity. The position vector includes the values for each of the variables in the problem. If the problem has two parameters, for instance, the particles will have position vectors with two dimensions. Each particle will then be able to move in an n-dimensional search space where n is the number of variables. To update the position of particles, the second vector (velocity) is considered. This vector defines the magnitude and direction of step size for each dimension and each particle independently.

The location of particles is updated in each step of optimization using the following equation:

$$\overrightarrow{X_i(t+1)} = \overrightarrow{X_i(t)} + \overrightarrow{V_i(t+1)} \tag{1}$$

where $\overrightarrow{X_i(t)}$ shows the position of ith particle at tth iteration and $V_i(t)$ shows the velocity of ith particle at tth iteration.

This equation shows that the position updating is simple and the main component is the velocity vector. The velocity vector is defined as follows:

$$\overrightarrow{V_i(t+1)} = w\overrightarrow{V_i(t)} + c_1 r_1 \left(\overrightarrow{P_i(t)} - \overrightarrow{x_i(t)} \right) + c_2 r_2 \left(\overrightarrow{G(t)} - \overrightarrow{x_i(t)} \right) \qquad (2)$$

where $\overrightarrow{X_i(t)}$ shows the position of ith particle at tth iteration, $\overrightarrow{V_i(t)}$ shows the velocity of ith particle at tth iteration, w is the inertial weight, c_1 shows the individual coefficient, c_2 signifies the social coefficient, r_1, r_2 are random numbers in $[0, 1]$, $\overrightarrow{P_i(t)}$ is the best solution obtained by the ith particle until tth iteration, and $\overrightarrow{G(t)}$ shows the best solution found by all particles (entire swarm) until tth iteration.

Equation 2 shows that the velocity vector is made of three components. The first component, $w\overrightarrow{V_i(t)}$ maintains the tendency towards the current velocity. This component is multiplied by an inertial parameter (w). The larger the value of this parameter, the higher the tendency to maintain the previous velocity. The second component, $c_1 r_1 \left(\overrightarrow{P_i(t)} - \overrightarrow{x_i(t)} \right)$ simulates the individual intelligence of a bird by memorizing and using the best solution obtained so far by each of the particles. The vector $\overrightarrow{P_i(t)}$ is updated in each iteration in case the ith particle finds a better solution. The impact of this component on the final value of the velocity can be increased or decreased by changing c_1. This parameter is multiplied by a random number in $[0, 1]$ to provide randomized behaviors since PSO is a stochastic optimization technique. Overall, the second component maintains a tendency towards the best solution that a particle found so far, the so called "personal best".

The third component, $c_2 r_2 \left(\overrightarrow{G(t)} - \overrightarrow{x_i(t)} \right)$ mimics the social intelligence of a flock of birds, in which the best solution obtained by all particles is saved in $\overrightarrow{G(t)}$ and used in this component. This means that considering the best solution found by the swarm gravitates all particles toward one point. The impact of this component can be tuned using c_2 as well.

With these three components, the next position of a particle can be defined. An example is shown in Fig. 1. This figure shows that each particle considers the previous

Fig. 1 In PSO, each particle considers the previous velocity, the personal best, and the global best to define the current velocity and update its position

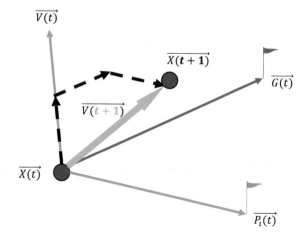

velocity, the personal best, and the global best to define the current velocity and update its position. The next position greatly depends on the random numbers generated using r_1 and r_2 as well. In Fig. 1, it is assumed that the particle considers 50% of the current velocity, 40% of the global best, and 40% of the personal best. In reality, however, c_1 and c_2 are multiplied by random numbers.

To see the area that a particle can move into, the values of w, c_1, and c_2 should be known. In the most well-regarded version of PSO, w linearly decreases from 0.9 to 0.4 proportional to the number of iterations. The parameters c_1, and c_2 are both set to 2 as well. Since it is difficult to show the possible locations of particles while varying the inertial weight, the snapshots of the possible locations when $w = 1$, $w = 0.5$, and $w = 0$ are shown in this subsection. The possible new positions of the particle with this configuration are visualized in Fig. 2.

The vectors toward the personal best and global best can be from 0 to double the size of their distance. This is due to the multiplication of both c_1 and c_2 by a random number in [0, 1]. Therefore, the range of these two parameters can be any number in the interval of [0, 2]. When the value is equal to 0, the particle does not consider the component. When the value is equal to 2, it considers twice the component. This is

Fig. 2 The impact of the inertial weight (w) on the velocity vector in PSO

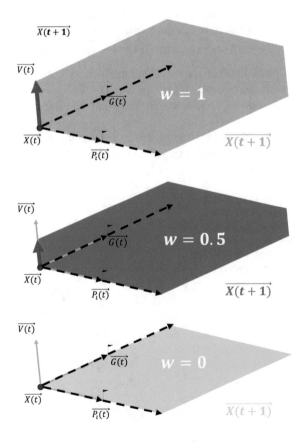

why the vectors are doubled in Fig. 2. The lower bound and upper bound of c_1 and c_2 do not change, so the maximum and minimum distance towards personal best and global best are equal as shown in Fig. 2. The inertial weight, however, is changed and its impact is evident in this figure.

Figure 2 shows that when the inertial weight is equal to 0, the next position of the particle is between its current location and the locations of personal and global bests. This causes exploitation and local search in the PSO model since the algorithm searches locally inside the area defined by the current position, personal best, and global best. In fact, the local search is purely done when c_1 and c_2 are both set to 1. An example is given in Fig. 3.

The exploration and global search increase proportionally to the values of all these parameters. This is shown in Fig. 4. It can be seen that the particle tends to go beyond the area between itself and the global or local bests. This leads to exploration and global search. It should be noted that the personal and global bests are updated constantly as well. So, the shaded area in Fig. 3, which shows the next position area, keeps changing as well.

The PSO algorithm uses these simple concepts to look for the global optimum of a given optimization problem. It starts with a random population of solutions. It then iteratively goes through the following steps until the end condition is met:

1. Calculating the objective value of all particles

Fig. 3 When $w = 0$, $c_1 = 1$ and $c_2 = 1$, the exploitation and local search is at the maximum level in PSO

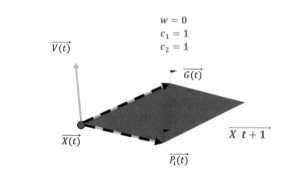

Fig. 4 The exploration and global search increases proportional to the values w, c_1 and c_2

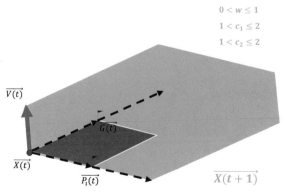

2. Updating w, c_1 and c_2. If c_1 and c_2 are constant, then only w is updated
3. Updating the personal best and the global best
4. Calculating the velocity vector for each particle using Eq. 1
5. Calculate the next position for each particle using Eq. 2

At the end of the optimization process, the best solution obtained by the entire swam will be returned as the final optimum for the optimization problem.

The PSO has been modified in a large number of works in the literature. There are also different variants of this algorithm too.

To solve binary problems, transfer functions have been widely integrated into PSO [9]. Such functions use the velocity values to flip a binary bit. The larger value of the velocity vector, the higher probability of flipping the binary bit [10, 11].

To solve multi-objective problems, several multi-objective variants of PSO have been proposed in the literature. The most popular technique was proposed by Coello et al. [12], in which an archive was employed to store and improve non-dominated solutions (Pareto optimal solutions) during the optimization process. In each iteration, one solution from the archive was chosen as the global best since all the solutions in the archive are considered a solution for a multi-objective optimization problem. To improve diversity of the solutions obtained, the archive was equipped with a grid mechanism [13]. There are also methods such as crowding distance to improve diversity as well [14]. Another popular technique is the use of non-dominated sorting in the PSO to rank and improve non-dominated solutions obtained so far [15]. There are also algorithms that aggregate objectives in the literature [16].

To handle constraints, there are many functions in the literature [17, 18]. Such functions are mostly integrated into the objective function to penalize the particles that violate the constraints. Some of these functions penalize the particles in a similar manner regardless of the level of violation. In this case, such functions can be called barrier functions. There are also functions that penalize particles proportional to their level of violation. Such functions are essential when solving problems with dominated infeasible regions since PSO deals with a large number of infeasible solutions in each iteration [19].

To solve problems with dynamically changing objectives, a dynamic version of PSO has been proposed in [20]. The main idea is to prevent PSO from converging towards a solution to track the changes in the objective function(s) [21]. To solve problems with uncertainties, robust PSOs [22, 23] and robust MOPSOs [24, 25] have been proposed in the literature as well.

3 Results

3.1 Exploration and Exploitation in PSO

In this subsection, the statements about the impact of the main controlling parameters (w, c_1 and c_2) on the performance of PSO have been investigated experimentally.

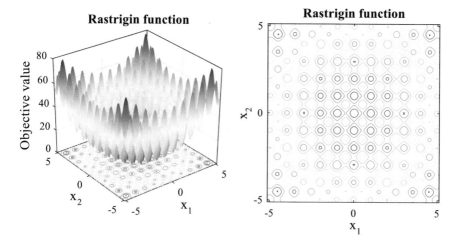

Fig. 5 Rastrigin test function used in the experiment

We test PSO while changing the parameters in a case study. Data visualization in a space with more than three dimensions is difficult to perform and analyze. Therefore, this section solves a 2-dimensional version of the Rastrigin test function as shown in Fig. 5.

The mathematical formulation of this test function is as follows:

$$f(\overrightarrow{x}) = An + \sum_{i=1}^{n}\left[x_i^2 - A\cos(2\pi x_i)\right] \tag{3}$$

where $A = 10$ and $x_i \in [-5.12, 5.12]$. The global optimum of this test function is right at the origin with the optimal value of 0.

Figure 5 shows that this function has a large number of local solutions and the search space is very challenging. To observe the exploratory and exploitative behaviour of PSO, the history of sampled points is visualized. If the solutions are widely spread around the search space, it is a good indication of high exploration. If the solutions scatter around just a region of the search space, it shows that the exploitation is high.

In the following sub-section, PSO is run with four particles and 500 iterations while changing the controlling parameters independently. To observe the exploratory and exploitative behaviours of PSO, the search history, convergence curve (GBEST in each iteration), average objective of all particles, fluctuations in the first dimension, and average distance between all particles are presented in multiple figures. Note that in the search history, the sampled points change their colours from cold to warm proportional to the number of iterations.

3.1.1 Investigating the Impact of the Inertial Weight (*w*)

In the first experiment, the value of the inertial weight is set to 1, 0.5, and 0. It also adaptively decreases from 0.9 to 0.2 proportional to the number of iterations.

The results in Fig. 6 show that when the inertial weight is equal to 0, the exploration is at the lowest level. This can be seen in the search history (the first column), average objective (the third column), fluctuation (the fourth column), and average distance

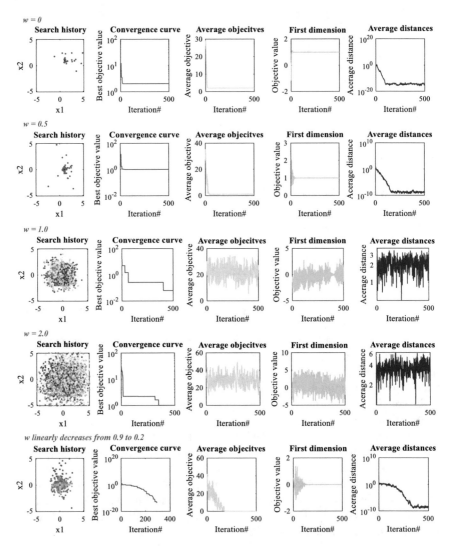

Fig. 6 The performance of PSO when the inertial weight is set to 1, 0.5, and 0. It also adaptively decreases from 0.9 to 0.2 proportionally to the number of iterations

between the particles (the fifth column). In addition, the exploitation is very high when using this value for the inertial weight. This configuration can be recommended for linear problems.

Figure 6 shows that exploration increases proportional to the value of the inertia weight. Considering the third and the fourth rows in this figure, it can be seen the search history covers wide areas. Since there are sampled points with warm color, it shows that particles keep exploring even in the final iteration. The fluctuations are substantial in the average objective, first dimension and the average distances too. These are all evidence of very high exploration when $w \geq 1$.

Neither of the above behaviours is enough to solve challenging problems. Both exploration and exploitation should be balanced. This can be done by adaptively changing the inertia weight. The last row in Fig. 6 shows that the algorithm first explores the search space (as the sampled points with cold colour show) and then exploits it. This allows finding an accurate estimation of the global optimum around iteration 300 as the convergence shows. The fluctuations in the average objectives, first dimension, and average distances are also reasonable, in that the changes taper away as the iteration count increases.

3.1.2 Investigating the Impact of the Cognitive Component (c_1)

In this subsection, different values (0, 1, 2, 3, and 4) are used for the cognitive constant in the PSO algorithm. The results are provided in Fig. 7. This figure shows that exploitation is very high when the cognitive constant is equal to 0. This is because all particles are pulled towards the global optimum since the impacts of the personal bests are skipped with this value. The average distance between particles is the most interesting curve to check here. The average distance between all particles is gradually decreasing, showing consistent movement of the particles towards each other and the global best.

On the other hand, the exploration increases proportional to the value of the cognitive constant as Fig. 7 shows. The reason why the algorithm finds the global optimum in most of the cases is due to the use of adaptive inertial weight in this experiment and the simplicity of the test function. For more challenging problems, it will be more difficult for PSO to find the global optimum. The exploration is very high when $c_1 = 3$ or $c_1 = 4$. In fact, the convergence and exploitation are interrupted when using $c_1 = 4$. This is because the particles almost exclusively gravitate toward their personal bests.

Figure 7 indicates that exploration and exploitation balances best at $c_2 = 2$, which is the reason why most works in the literature uses this value for the cognitive constant.

3.1.3 Investigating the Impact of the Social Component (c_2)

In this subsection, different values (0, 1, 2, 3, and 4) are used for the social constant in the PSO algorithm. The results are provided in Fig. 8.

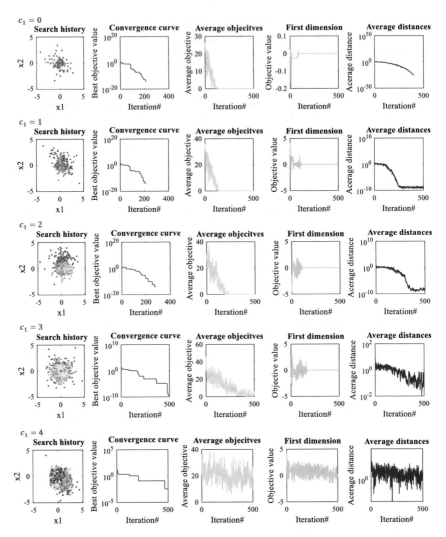

Fig. 7 The performance of PSO when the cognitive constant is set to 0, 1, 2, 3, and 4

An interesting pattern can be observed when $c_2 = 0$, in which four particles converged towards independent points. This is due to the fact that the global best is not considered when $c_2 = 0$, so each particle searches around its personal best independently. In this case, there is no systematic convergence towards one solution at the end.

As the social constant increases, particles start to use the global best, and the entire swarm is more directed. A reasonable balance between local and global search is observed when using $c_2 = 2$. For larger values, the impact from the global best dominates, and the PSO algorithm behaves more randomly.

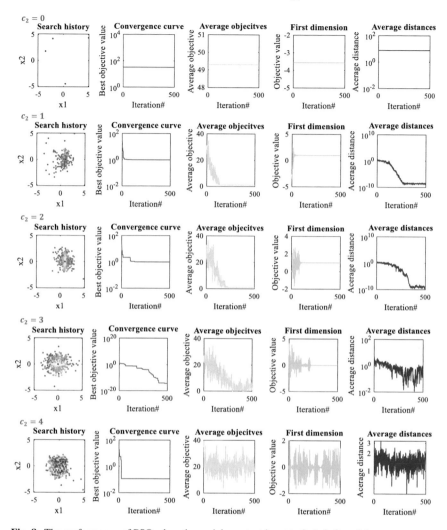

Fig. 8 The performance of PSO when the social constant is set to 0, 1, 2, 3, and 4

Overall, the results so far have shown that the best configuration is to linearly decrease the inertia weight while setting c_1 and c_2 to 2. This configuration will be used when solving the real-world problem in the next section.

3.2   2D Airfoil Design Using PSO

The problem of 2D airfoil design deals with designing the cross section of an aircraft wing considering two objectives: maximizing lift and minimizing drag. Since the main focus of this book and chapter is on single-objective optimization, the problem is solved considering the latter objective.

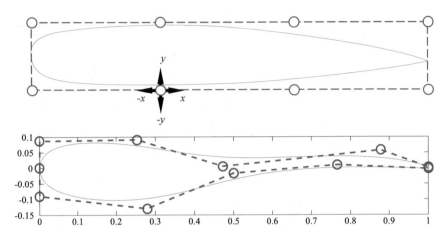

Fig. 9 B-spline method with controlling points to define the shape of an airfoil

Several variants of this problem have been solved in the literature. They mostly use different methods and parameters to define the shape of the airfoil. This subsection uses the B-spline method as shown in Fig. 9. This figure shows that there are eight controlling parameters that can be moved in two dimensions. As such, the total number of variables is equal to 16.

The problem of 2D airfoil design is formulated as follows:

$$\text{Maximize: } C_l\left(\overrightarrow{X}, \overrightarrow{Y}\right) \tag{4}$$

$$\text{Minimize: } C_d\left(\overrightarrow{X}, \overrightarrow{Y}\right) \tag{5}$$

$$\text{Subject to: } -1 \leq \overrightarrow{X}, \overrightarrow{Y} \leq 1 \tag{6}$$

where C_d is the coefficient of drag, $\overrightarrow{X} = \{x_1, x_2, ..., x_8\}$ and $\overrightarrow{Y} = \{y_1, y_2, ..., y_8\}$

There are a large number of constraints in this problem. Since the details of the problem is outside the scope of this book, interested readers may refer to [26]. To calculate the coefficient of drag a free-ware called XFoil is employed [27].

The above problem is multi-objective, but the experiments described in this chapter optimizes each of the objectives independently. In addition, $\frac{C_l}{C_d}$ is maximized, which should have a tendency to maximize the lift and minimize the drag. All three versions of this problem are solved by the PSO algorithm using 10 particles and 150 iterations. PSO is required to stop in case of exceeding the maximum number of iterations or not getting a better result in more than 100 iterations. The results are shown in Fig. 10. This figure shows that the objective function (C_l/C_d) is increased

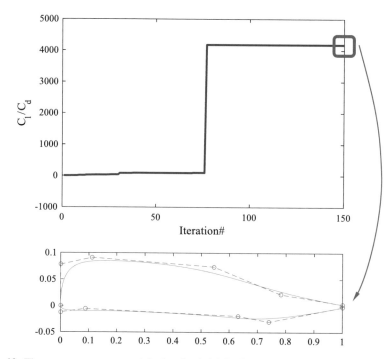

Fig. 10 The convergence curve and final optimal airfoil when maximizing C_l/C_d

significantly only once. This shows that PSO is not better than a random search in this problem, which is due to the large number of infeasible particles in each iteration that hinder the search mechanism of PSO. The convergence curve shows the improvement in the objective value, yet it does not show the shapes of airfoils. To see how the random population of PSO looks like, Fig. 11 is given. This figure shows 10 particles in the initial population.

It can be seen in this figure that almost all of the initial shapes are not smooth and do not provide lift. Most of them even violate constraints, so they are considered infeasible. Two popular methods to alleviate these drawbacks are limiting the range of parameters and starting with initial feasible population. This is done at the end of this chapter.

This chapter also optimizes both of the objectives independently. The same experimental settings compared to the previous experiments are used to find an optimal design for the 2D airfoil. The convergence of PSO and the best solution found are shown in Figs. 12 and 13. The convergence of the PSO algorithm when maximizing C_l is similar to that shown in Fig. 10 when maximizing C_l/C_d. There is a big incline at a point during the iterations and there is no significant improvement afterwards. In fact, there is no improvement for 100 iterations, which is the reason why the

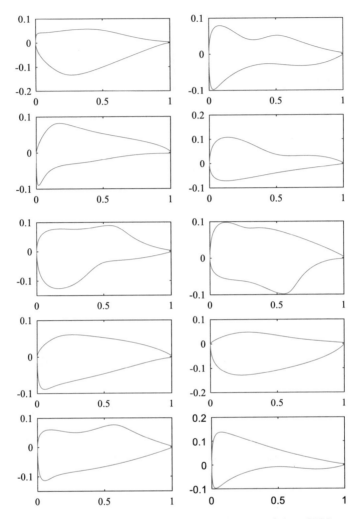

Fig. 11 Ten initial airfoils repressed by each particle in the first population of PSO

algorithm stopped before hitting the 150th iteration. This is due to the same reason mentioned when maximizing C_l/C_d.

The convergence of the PSO algorithm when minimizing C_d is different from those on other objectives. Figure 13 shows that the best solution keeps improving over the course of iteration. Since the algorithm has not been changed, this shows that the search space of the drag objective function is less challenging than that of the lift. Any airoilf shape can give drag, but not all aoifoil shapes provide lift. The results on the other two objectives show that once the lift is involved in the objective function, it becomes more difficult for the algorithm to maintain a consistent improvement

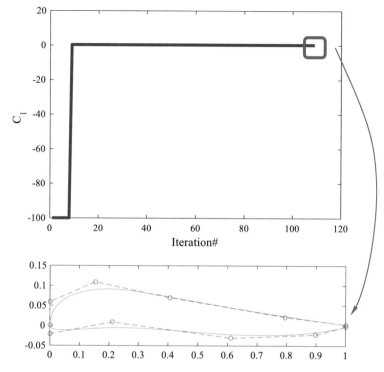

Fig. 12 The convergence curve and final optimal airfoil when maximizing C_l

and steady convergence rate. This is due to the large number of infeasible solutions in each iteration. The accuracy of PSO cannot be improved with fine tuning since the main problem is the failure of PSO mechanisms when there are not at least two feasible solutions in the population.

To see how much the performance of PSO can be improved when considering the lift objective function, the range of variables are bounded significantly and some feasible solutions are added in the first population. The results are shown in Fig. 14. This figure shows that the PSO algorithm shows consistent improvement for such a highly-constraint problem when incorporating the changes.

Taken together, the results of this chapter showed that the PSO algorithm is very beneficial in finding an optimal shape for a 2D airfoil. The experiments on the test functions also showed that tuning the parameters of PSO should be done carefully since using inappropriate values might significantly degrade its performance and change the exploratory and exploitative behaviours.

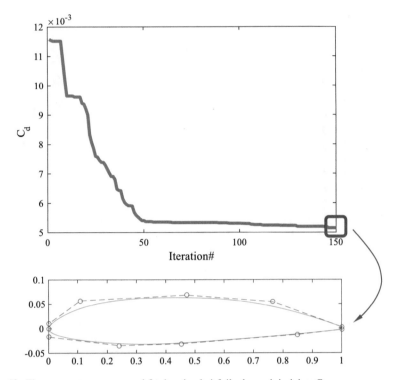

Fig. 13 The convergence curve and final optimal airfoil when minimizing C_d

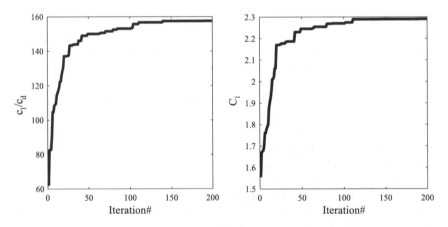

Fig. 14 The convergence curve and final optimal airfoil (maximizing C_l or C_l/C_d) when starting with some feasible solutions and limiting the parameters' bounds

4 Conclusion

This chapter first presented the inspirations and mathematical models of the PSO algorithm. The impact of the parameters on the particle movement and position updating equations was then discussed theoretically. After providing a brief literature review covering different variants, the PSO algorithm was applied to challenging benchmark functions. A large number of figures were used to observe the performance of the PSO algorithm including the search history, convergence curve, average objective of all particles, fluctuation in one variable, and average distance between all particles. The experiments were conducted while changing the values for the main controlling parameters of PSO: inertial weight (w), cognitive constant (c_1), and social constant (c_2).

It was observed that the best balance between exploration and exploitation can be achieved when linearly decreasing the inertial weight. Also, the best value to achieve a similar balance is to set both cognitive and social component to 2. After extensive experiments on the benchmark functions, the chapter employs the PSO algorithm to find an optimal shape for a 2D aifoil considering three different objectives. The results showed that PSO is able to find an optimal design to maximize or minimize any of the objectives.

References

1. Faris, H., Aljarah, I., Al-Betar, M. A., & Mirjalili, S. (2017). Grey wolf optimizer: A review of recent variants and applications. *Neural Computing and Applications*, 1–23.
2. Kirkpatrick, S., Gelatt, C. D.,& Vecchi, M. P. (1983). Optimization by simulated annealing. *Science, 220*(4598), 671–680.
3. Holland, J. H. (1992). Genetic algorithms. *Scientific American, 267*(1), 66–73.
4. Storn, R., & Price, K. (1997). Differential evolutiona simple and efficient heuristic for global optimization over continuous spaces. *Journal of Global Optimization, 11*(4), 341–359.
5. Dorigo, M., & Birattari, M. (2011). Ant colony optimization. In *Encyclopedia of machine learning* (pp. 36–39). Boston: Springer.
6. Kennedy, J. (2011). Particle swarm optimization. In *Encyclopedia of machine learning* (pp. 760–766). Boston: Springer.
7. Chitsaz, H., & Aminisharifabad, M. (2015). Exact learning of rna energy parameters from structure. *Journal of Computational Biology, 22*(6), 463–473.
8. Aminisharifabad, M., Yang, Q., & Wu, X. (2018). A penalized autologistic regression with application for modeling the microstructure of dual-phase high strength steel. *Journal of Quality Technology* (in-press).
9. Clerc, M. (2004). Discrete particle swarm optimization, illustrated by the traveling salesman problem. In *New optimization techniques in engineering* (pp. 219–239). Berlin: Springer.
10. Mirjalili, S., & Lewis, A. (2013). S-shaped versus V-shaped transfer functions for binary particle swarm optimization. *Swarm and Evolutionary Computation, 9*, 1–14.
11. Saremi, S., Mirjalili, S., & Lewis, A. (2015). How important is a transfer function in discrete heuristic algorithms. *Neural Computing and Applications, 26*(3), 625–640.
12. Coello, C. A. C., Pulido, G. T., & Lechuga, M. S. (2004). Handling multiple objectives with particle swarm optimization. *IEEE Transactions on Evolutionary Computation, 8*(3), 256–279.

13. Mostaghim, S., & Teich, J. (2003). Strategies for finding good local guides in multi-objective particle swarm optimization (MOPSO). In *Proceedings of the 2003 IEEE Swarm Intelligence Symposium, 2003. SIS 2003* (pp. 26–33). IEEE.
14. Sierra, M. R., & Coello, C. A. C. (2005). Improving PSO-based multi-objective optimization using crowding, mutation and-dominance. In *International Conference on Evolutionary Multi-Criterion Optimization* (pp. 505–519). Berlin: Springer.
15. Li, X. (2003). A non-dominated sorting particle swarm optimizer for multiobjective optimization. In *Genetic and Evolutionary Computation Conference* (pp. 37–48)
16. Reyes-Sierra, M., & Coello, C. C. (2006). Multi-objective particle swarm optimizers: A survey of the state-of-the-art. *International Journal of Computational Intelligence Research, 2*(3), 287–308.
17. Parsopoulos, K. E., & Vrahatis, M. N. (2002). Particle swarm optimization method for constrained optimization problems. *Intelligent TechnologiesTheory and Application: New Trends in Intelligent Technologies, 76*(1), 214–220.
18. Pulido, G. T., & Coello, C. A. C. (2004). A constraint-handling mechanism for particle swarm optimization. In *IEEE congress on evolutionary computation* (Vol. 2, pp. 1396–1403).
19. Coello, C. A. C. (2002). Theoretical and numerical constraint-handling techniques used with evolutionary algorithms: A survey of the state of the art. *Computer Methods in Applied Mechanics and Engineering, 191*(11–12), 1245–1287.
20. Eberhart, R. C., & Shi, Y. (2001). Tracking and optimizing dynamic systems with particle swarms. In *Proceedings of the 2001 Congress on Evolutionary Computation, 2001* (Vol. 1, pp. 94–100). IEEE.
21. Blackwell, T., & Branke, J. (2004). Multi-swarm optimization in dynamic environments. In *Workshops on Applications of Evolutionary Computation* (pp. 489–500). Berlin: Springer.
22. Luan, F., Choi, J. H., & Jung, H. K. (2012). A particle swarm optimization algorithm with novel expected fitness evaluation for robust optimization problems. *IEEE Transactions on Magnetics, 48*(2), 331–334.
23. Mirjalili, S., Lewis, A., & Mostaghim, S. (2015). Confidence measure: A novel metric for robust meta-heuristic optimisation algorithms. *Information Sciences, 317*, 114–142.
24. Mirjalili, S., Lewis, A., & Dong, J. S. (2018). Confidence-based robust optimisation using multi-objective meta-heuristics. *Swarm and Evolutionary Computation*.
25. Ono, S., Yoshitake, Y., & Nakayama, S. (2009). Robust optimization using multi-objective particle swarm optimization. *Artificial Life and Robotics, 14*(2), 174.
26. Mirjalili, S., Lewis, A., & Mirjalili, S A M. (2015) Multi-objective optimisation of marine propellers. *Procedia Computer Science, 51*, 2247–2256.
27. Drela, M. (1989). XFOIL: An analysis and design system for low Reynolds number airfoils. In *Low Reynolds number aerodynamics* (pp. 1–12). Springer, Berlin, Heidelberg.

Salp Swarm Algorithm: Theory, Literature Review, and Application in Extreme Learning Machines

Hossam Faris, Seyedali Mirjalili, Ibrahim Aljarah, Majdi Mafarja
and Ali Asghar Heidari

Abstract Salp Swarm Algorithm (SSA) is a recent metaheuristic inspired by the swarming behavior of salps in oceans. SSA has demonstrated its efficiency in various applications since its proposal. In this chapter, the algorithm, its operators, and some of the remarkable works that utilized this algorithm are presented. Moreover, the application of SSA in optimizing the Extreme Learning Machine (ELM) is investigated to improve its accuracy and overcome the shortcomings of its conventional training method. For verification, the algorithm is tested on 10 benchmark datasets and compared to two other well-known training methods. Comparison results show that SSA based training methods outperforms other methods in terms of accuracy and is very competitive in terms of prediction stability.

1 Introduction

Artificial Neural Networks (ANNs) are mathematical models that are widely applied in machine learning for supervised and unsupervised tasks [4, 26, 29, 30, 41]. ANNs can be distinguished based on their architecture and the learning algorithm.

H. Faris · I. Aljarah
King Abdullah II School for Information Technology, The University of Jordan, Amman, Jordan
e-mail: hossam.faris@ju.edu.jo

I. Aljarah
e-mail: i.aljarah@ju.edu.jo

S. Mirjalili (✉)
Institute of Integrated and Intelligent Systems, Griffith University, Nathan, Brisbane, QLD 4111, Australia
e-mail: seyedali.mirjalili@griffithuni.edu.au

M. Mafarja
Department of Computer Science, Faculty of Engineering and Technology, Birzeit University, PoBox 14, Birzeit, Palestine
e-mail: mmafarja@birzeit.edu

A. A. Heidari
School of Surveying and Geospatial Engineering, University of Tehran, Tehran, Iran
e-mail: as_heidari@ut.ac.ir

© Springer Nature Switzerland AG 2020
S. Mirjalili et al. (eds.), *Nature-Inspired Optimizers*, Studies in Computational
Intelligence 811, https://doi.org/10.1007/978-3-030-12127-3_11

185

The architecture of the network refers to the way that their basic processing elements are connected and distributed over a number of layers. While the learning algorithm is responsible for optimizing the connection weights of the network in order to minimize or maximize some predefines criteria [5, 28, 31].

The most popular type of ANN in the literature is the Single Hidden Layer Feedforward Networks (SLFN). SLFN is typically trained by the Back-propagation algorithm, which is a gradient descent training method. In spite of its popularity, this algorithm has some drawbacks such as the high sensitivity to the initial weights, high probability to be trapped in a local minima, slow convergence, and the need to carefully set its parameters like the learning rate and momentum [19, 34, 35, 66].

In an attempt to overcome the aforementioned problems, Extreme Learning Machine (ELM) was proposed by Huang et al. in [45]. ELM is extremely fast training method for SLFN which starts by initializing the input weights and hidden biases randomly, then it calculates the output weights analytically using Moore-Penrose (MP) generalized inverse. ELMs have several advantages over the traditional gradient descent based neural networks. ELM advantages include: (1) they are extremely fast to train, (2) no need for human intervention and extra effort for tuning some parameters like the learning rate and momentum such as gradient descent algorithms. Due to these advantages, ELM became very popular in the last decade with wide range of successful applications [23].

Although ELM has promising generalization performance and it eliminates the need for tuning some parameters like in the Back-propagation case, it showed tendency to toward the need for higher number of processing elements in the hidden layer [18]. In addition, the random initialization of the input weights in ELM could affect its generalization performance [22, 61].

In the last few years, different approaches were proposed to improve the performance of ELM networks. One noticeable research line is the deployment of evolutionary and swarm based intelligence algorithms to optimize the input weights and biases of ELM. Swarm intelligence techniques are a hot subject of studies to be used in many trending applications such as machine learning [3, 24, 55], global mathematical optimization [10, 40–42, 44], spatial problems [38, 39, 43], spam and intrusion detection systems [7, 9, 27, 31], feature selection [11, 24, 25, 50–58], clustering analysis [2, 6, 8, 67], optimizing neural networks [4, 5, 26, 29], link prediction [14], software effort estimation [33], and bio-informatics [12, 17, 37]. Moreover, swarm intelligence techniques are used in training ELM such as Differential Evolution (DE) and Particle Swarm Optimization (PSO) [36, 64, 69, 70].

In this work, a new training method for ELM is proposed based on a recent nature-inspired optimization algorithm called Salp Swarm Algorithm (SSA) [59]. SSA is deployed to evolve the input connection weights and hidden biases of the single hidden layer feedforward network of ELM. We will refer to this algorithm as SSA-ELM. SSA has shown recently very promising results when applied for complex engineering machine learning problems [59]. This has ignited a motivation for exploring the efficiency of its operator in evolving ELM networks to overcome their shortcomings.

2 Extreme Learning Machines

ELM is a learning framework for single hidden layer feedforward neural networks (SLFN). ELM was introduced by Haung in [46]. The general idea of ELM is to randomly generate the connection weights between the input layer and the hidden layer then the weights that connect the hidden layer to the output layer are analytically computed. The architecture of ELM is shown in Figure.

For N arbitrary distinct samples shown by (x_i, t_i), where $x_i = [x_{i1}, x_{i2}, , x_{iN}]^T \in \mathbb{R}^n$ and $t_i = [t_{i1}, t_{i2}, , t_{iN}]^T \in \mathbb{R}^m$, a standard SLFN with an activation function $g(x)$ and $\tilde{N}$ neurons in hidden layer can be mathematically represented as in Eq. (1) [45]:

$$\sum_{i=1}^{\tilde{N}} \beta_i g(\omega_i \cdot x_j + b_i) = o_j, j = 1, \dots, N \tag{1}$$

Where b_i is the threshold of the ith hidden neuron, $w_i = [w_{i1}, w_{i2}, \dots, w_{in}]^T$ is the weight vector that connects the ith hidden neuron with the input neurons, $\beta_i = [\beta_{i1}, \beta_{i2}, \dots, \beta_{im}]^T$ is the vector of weights that connects the ith hidden neuron with output neurons [45].

The conventional SLFNs with $\tilde{N}$ hidden neurons and activation function $g(x)$ are capable of approximating the initial N samples with zero error means that $\sum_{i=1}^{\tilde{N}} \|o_j - t_j\| = 0$, i.e., we have w_i, β_i, and b_i such that [45]:

$$\sum_{i=1}^{\tilde{N}} \beta_i g(\omega_i \cdot x_j + b_i) = t_j, j = 1, \dots, N \tag{2}$$

The aforementioned N rules can be expressed as given in Eq. (3):

$$H\beta = T \tag{3}$$

where

$$H(w_1, \dots, w_N, b_1, \dots, b_N, x_1, \dots, x_N) =$$
$$\begin{bmatrix} g(w_1.x_1 + b_1) & \dots & g(w_{\tilde{N}}.x_1 + b_{\tilde{N}}) \\ \vdots & & \vdots \\ g(w_1.x_N + b_1) & \dots & g(w_{\tilde{N}}.x_N + b_{\tilde{N}}) \end{bmatrix}_{N \times \tilde{N}} \tag{4}$$

$$\beta = \begin{bmatrix} \beta_1^T \\ \vdots \\ \beta_{\tilde{N}}^T \end{bmatrix}_{\tilde{N} \times m}, T = \begin{bmatrix} t_1^T \\ \vdots \\ t_{\tilde{N}}^T \end{bmatrix}_{N \times m} \tag{5}$$

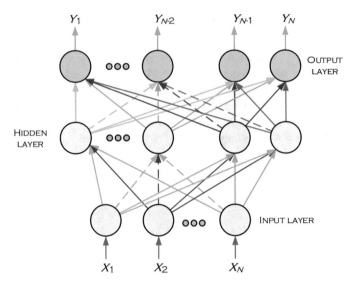

Fig. 1 Overall structure of ELM

where H is the hidden layer output matrix, the ith column of H is the ith hidden output vector of neuron with regard to $x_1, x_2, \ldots, x_N$ [45]. Figure 1 shows the overall structure of the ELM.

3 Salp Swarm Algorithm

The SSA is a recent nature-inspired optimizer proposed by Mirjalili et al. [59] in 2017. The purpose of SSA is to develop a population-based optimizer by mimicking the swarm behavior of salps in nature [1]. The performance of the original SSA as an ELM trainer has not been investigated to date. SSA algorithm reveals satisfactory diversification and intensification propensities that make it appealing for evolving ELM training tasks. The unique advantages of SSA cannot be obtained by using some traditional optimizers such as PSO, GWO, and GSA techniques. The SSA can be considered as a capable, flexible, simple, and easy to be understood and utilized in parallel and serial modes. Furthermore, it has only one adaptively decreasing parameter to make a fine balance between the diversification and intensification inclinations. In order to avoid immature convergence to local optima (LO), the position vectors of salps are gradually updated considering other salps in a dynamic crowd of agents. The dynamic movements of salps enhance the searching capabilities of the SSA in escaping from LO and immature convergence drawbacks. It also keeps the elite salp found so far to guide other members of swarm towards better areas of the feature space.

The SSA has an iterative nature. Hence, it iteratively generates and evolves some random individuals (i.e., salps) inside the bounding box of the problem. Then, all

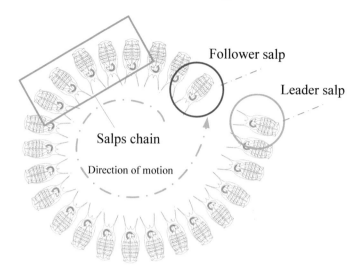

Fig. 2 Illustration of salp's chain and the concept of leader and follower

salps should update their location vectors (leader of chain and followers). The leader salp will attack in the direction of a food source (F), while all followers can move towards the rest of salps (and leader directly or indirectly) [59]. An illustration of salp chain is shown in Fig. 2.

The population of salps X consists of N agents with d-dimensions. Hence, it can be exposed by a $N \times d$-dimensional matrix, as described in Eq. (6):

$$X_i = \begin{bmatrix} x_1^1 & x_2^1 & \cdots & x_d^1 \\ x_1^2 & x_2^2 & \cdots & x_d^2 \\ \vdots & \vdots & \cdots & \vdots \\ x_1^N & x_2^N & \cdots & x_d^N \end{bmatrix} \tag{6}$$

In SSA, the position of leader is calculated by Eq. (7):

$$x_j^1 = \begin{cases} F_j + c_1 \left((ub_j - lb_j) c_2 + lb_j \right) & c_3 \geq 0.5 \\ F_j - c_1 \left((ub_j - lb_j) c_2 + lb_j \right) & c_3 < 0.5 \end{cases} \tag{7}$$

where x_j^1 denotes the leader's location and F_j reveals the position vector of food source in the jth dimension, ub_j shows the superior limit of jth dimension, and lb_j represents the inferior limit of jth dimension, c_2 and c_3 are random values inside [0, 1], c_1 is the main parameter of algorithm expressed as in Eq. (8):

$$c_1 = 2e^{-(\frac{4t}{T_{max}})^2} \tag{8}$$

where t is the iteration, while T_{max} shows the maximum number of iterations. By increasing iteration count, this parameter decreases. As a result, it can manage to put more emphasize on the diversification inclination on initial stages and put more emphasize on intensification tendency in last steps of optimization. The location of followers are adjusted by Eq. (9):

$$x_j^i = \frac{x_j^i + x_j^{i-1}}{2} \qquad (9)$$

where $i \geq 2$ and x_j^i is the location of the ith follower salp at the jth dimension. The pseudo-code of SSA is expressed in Algorithm 1.

Algorithm 1 Pseudo-code of the SSA

Initialize the swarm $x_i (i = 1, 2, \ldots, n)$
while (end condition is not met) **do**
   Obtain the fitness of all salps
   Set **F** as the leader salp
   Update c_1 by Eq. (11.8)
   for (every salp (x_i)) **do**
     if $(i == 1)$ **then**
       Update the position of leader by Eq. (11.7)
     else
       Update the position of followers by Eq. (11.9)
   Update the population using the upper and lower limits of variables
   Return back salps that violated the bounding restrictions.
Return **F**

Based on Algorithm 1, we see that SSA first initiates all salps, randomly. Then, it evaluates all salps to select the fittest salp of swarm F. The leader will be followed by the chain of salps as shown in Fig. 2. Meanwhile, the variable c_1 is adjusted by Eq. (8). Equation (7) assists SSA in updating the location of leader, while Eq. (9) can update the position vector of the follower salps. Until the last iteration, all these steps except the initialization step have to be repeated.

4 Literature Review

Since its release, SSA has been applied in various applications in which it showed its efficiency and competitiveness. In this section, noticeable recent works on SSA and their main results are reviewed.

One of the key works on SSA was presented in [24] by Faris et. al to deal with feature selection (FS) tasks. The work proposed two wrapper feature selection approaches based on binary versions of SSA (BSSA). In the first approach, eight transfer functions are used to transform the continuous version of SSA to binary.

In the second approach, a crossover operator replaced the average operator to deepen the exploration tendency in the original algorithm. The proposed approaches were tested using 22 benchmark datasets and the results were compared with other FS methods. The results showed that the proposed SSA-based approaches can significantly outperform other well-established approaches on the majority of the datasets. Another work on FS was conducted by Sayed et al. in [65]. The authors presented a new chaos-based SSA (CSSA) to deal with FS datasets. Simulation results revealed that the CSSA can be regarded as a good optimizer compared to some previous methods. As a specific FS application, SSA was presented in combination with k-nearest neighbors (k-NN) to choose a small number of features and achieve higher classification accuracies in predicting chemical compound activities [47]. Comparison results showed the efficiency of SSA compared to other well-known optimizers.

Ismael et al. [48] used basic SSA to tackle the task of choosing the best conductor in a real-life radial distribution system in Egypt. SSA was applied also for tuning of power system stabilizer in a multi-machine power system [20]. The presented experiments showed that SSA outperformed other intelligent techniques. In [62], SSA is used to optimize the gains and the parameters of the fractional order proportional-integral derivative controller based on an integral time absolute error as objective function. Another application in electric engineering, the authors in [13] proposed the application of SSA for optimizing the sizing of a CMOS differential amplifier and the comparator circuit. Their conducted experiments showed that the performance of the proposed SSA with CMOS analog IC designs are better than other reported studies.

Other engineering problems include optimizing load frequency control using SSA in managing the active power of an isolated renewable microgrid [15], extracting the parameters of polarization curves of polymer exchange membrane fuel cells model [21], design of PID-Fuzzy control for seismic exited structural system against earthquake [16], and solving economic load dispatch problem [63].

SSA was also deployed in environmental applications. In [71], SSA was utilized to optimize the hyper-parameters of the least squares support sector machine for forecasting energy-related CO_2 emissions. The results showed the proposed model is superior and has the potential to improve the forecasting accuracy and reliability of CO_2 emissions.

All these works show that the SSA has a high capability in managing the fine exploration and exploitation tendencies especially for FS tasks. This method can show satisfactory convergence trends, relatively deep exploration and exploitation of the feature space, and flexibility in detecting near-optimal solutions.

5 Proposed SSA-ELM Approach

In this section, we describe the design and procedure of the proposed SSA-ELM training algorithm. In SSA-ELM, the operators of SSA are utilized to evolve ELM networks, where each salp represents a candidate ELM network. To capture this

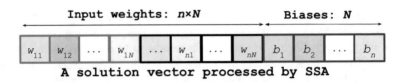

Input weights: n×N

Fig. 3 Structure design of the salp used in the proposed SSA-ELM

representation, the salp is designed to hold the parameters of the network that we want to optimize. In this case, these parameters are the synaptic weights that connect the input layer to the output layer and the biases on the hidden neurons. Therefore, the length (L) of each salp is $I \times N + N$, where I is the number of input variables. The structure design of the salp is represented as shown in Fig. 3.

In order to assess the quality of the generated solutions, a fitness function should be used. In this work we use one of the most common fitness functions which minimization of misclassification rate. This rate can be expressed as given in Eq. 10.

$$f = min(1 - Accuracy) \tag{10}$$

Where the $Accuracy$ is a total number of correctly classified instances over the total number of instances in the training set and it can be calculated as given in Eq. 11.

$$Accuracy = \frac{\sum_{i=1}^{n} \sum_{j=1}^{c} f(i, j) C(i, j)}{n} \tag{11}$$

Where n is the number of instances in the training set, and c is the number of classes. $f(i, j)$ is an indicator function that returns 0 or 1. $f(i, j)$ is 1 if the instance i is of class j. $C(i, j)$ is 1 iff the predicted class of instance i is j, otherwise 0.

6 Experiments and Results

The proposed SSA-ELM approach is evaluated based on 10 benchmark datasets drawn from the University of California at Irvine (UCI) Machine Learning Repository [49]. The datasets are described in terms of number of instances and features in Table 1. The datasets are varied in their specifications to form different levels of complexity for the experimented algorithms. All datasets are normalized using Min-Max normalization method in the range [0, 1] [3, 32, 55, 67].

For verification, SSA-ELM is compared with other two metaheuristic based ELM models commonly used in the literature which are PSO-ELM and DE-ELM. In addition SSA-ELM is compared with the basic ELM network. The parameters of

Table 1 Datasets specifications

Dataset	Instances	Features
Australian	690	14
Blood	748	4
Breast cancer	699	8
Chess	3196	36
German	1000	24
Habitit	155	10
Ring	7400	20
Tictactoe	958	9
Titanic	2201	3
Twonorm	7400	20

Table 2 Parameters of algorithms

Algorithm	Parameter	Value
PSO	Inertia factor	0.2–0.9
	c_1	2
	c_2	2
DE	CR Crossover	0.8
	F Scaling	1

PSO, DE and SSA are set as listed in Table 2. The number of hidden neurons in all algorithms is set as $2 \times F + 1$ where F is the number input features in the dataset. This rule is commonly used in the literature for SLFNs [4, 28, 60, 68]. The number of iterations and population/swarm size in all algorithms is set to 50.

In order to obtain credible results, each algorithm is trained and tested using 10-folds cross validation. Then, the average of the accuracy rates and the standard deviations are calculated and reported along with the best accuracy rates.

The convergence curves of PSO-ELM, DE-ELM and SSA-ELM for all datasets are shown in Figs. 4 and 5. It can be noticed that SSA-ELM has a fast convergence than the other two algorithms in most of the datasets. In Table 3, we list the average accuracy results, standard deviations and the best obtained accuracy rates. Inspecting the results, it can be seen that the SSA-ELM obtained the highest average of accuracies in 9 datasets with a noticeable difference in some datasets like Australian, German, Habitit and Tictactoe datasets. Considering the values of standard deviations, SSA-ELM shows the smallest values or very competitive to the other algorithms. This indicates the stability and robustness of the algorithm. In addition, SSA-ELM hit the highest accuracy rates in 6 datasets.

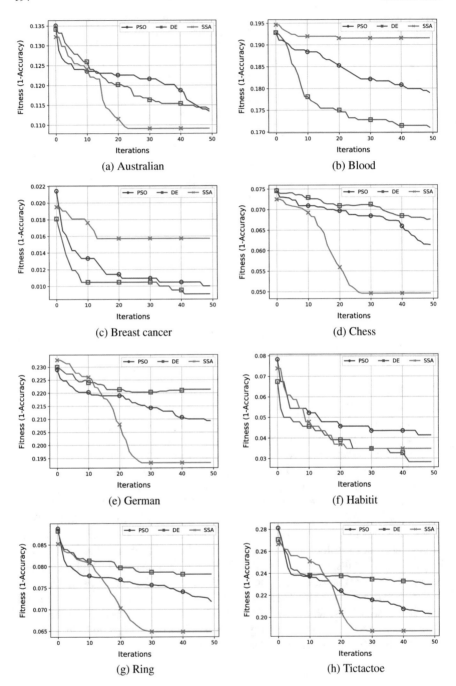

Fig. 4 Convergence curves for SSA, DE and PSO

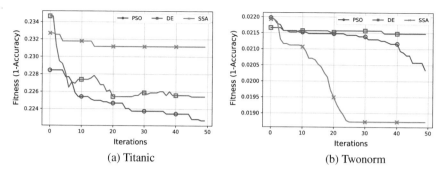

(a) Titanic   (b) Twonorm

Fig. 5 Convergence curves for SSA, DE and PSO

Table 3 Average and best accuracy rates of ELM networks

Dataset	PSO avg±stdv [best]	DE avg±stdv [best]	SSA avg±stdv [best]	ELM avg±stdv [best]
Australian	85.74 ± 3.54 [92.65]	83.68 ± 3.83 [92.65]	**88.24 ± 3.1** [92.65]	86.03 ± 2.96 [89.71]
Blood	77.97 ± 4.03 [83.78]	79.32 ± 3.06 [82.43]	**79.86 ± 3.02** [83.78]	78.65 ± 3.24 [83.78]
Breast	96.09 ± 2.98 [100.0]	95.94 ± 1.65 [98.55]	**96.52 ± 2.83** [100.0]	95.94± 2.44 [98.55]
Chess	93.02 ± 1.91 [95.91]	93.18 ± 1.63 [95.91]	**93.27 ± 1.19** [94.97]	92.61± 1.43 [94.34]
German	74.3 ± 4.4 [82.0]	73.5 ± 3.21 [77.0]	**76.6 ± 2.12** [81.0]	74.0± 3.62 [81.0]
Habitit	79.33 ± 8.58 [93.33]	82.0 ± 6.32 [93.33]	**82.0 ± 9.45** [100.0]	76.67± 9.56 [86.67]
Ring	91.69 ± 1.11 [93.23]	91.6 ± 1.05 [92.96]	**92.02 ± 1.27** [93.64]	88.57 ± 2.36 [91.07]
Tictactoe	73.37 ± 3.06 [77.9]	72.42 ± 4.98 [78.95]	**75.47 ± 3.44** [80.0]	68.32 ± 4.72 [74.74]
Titanic	78.18 ± 1.87 [81.36]	78.55 ± 2.39 [84.09]	**78.73 ± 1.78** [81.36]	77.23 ± 2.54 [79.09]
Twonorm	97.4 ± 0.65 [98.51]	**97.71 ± 0.43** [98.38]	97.62 ± 0.71 [98.38]	97.4 ± 0.55 [98.24]

7 Conclusion and Future Directions

In this chapter we reviewed one of the recent promising metaheuristic algorithms called Salp Swarm Algorithm (SSA), which is inspired by the swarming behavior of salps in seas. Moreover, we investigated the efficiency of SSA in training extreme learning machine (ELM). Specifically, SSA is used to optimize the input weights and biases of the hidden layer in ELM. The SSA based training method is experimented using 10 benchmark datasets and compared to two other well-knows methods which are Particle Swarm Optimization and Differential Evolution. Evaluation results show that the proposed methods outperform others in terms of classification accuracy. For future work, it is planned to study the potential of SSA in training other types of neural networks like Radial Basis Function networks (RBFN), Kernel ELM (KELM) and Regularized ELM.

References

1. Abbassi, R., Abbassi, A., Heidari, A. A., & Mirjalili, S. (2019). An efficient salp swarm-inspired algorithm for parameters identification of photovoltaic cell models. *Energy Conversion and Management, 179*, 362–372.
2. Al-Madi, N., Aljarah, I., & Ludwig, S. (2014). Parallel glowworm swarm optimization clustering algorithm based on mapreduce. In *IEEE Symposium Series on Computational Intelligence (IEEE SSCI 2014)*. IEEE Xplore Digital Library.
3. Aljarah, I., AlaM, A. Z., Faris, H., Hassonah, M. A., Mirjalili, S., & Saadeh, H. (2018). Simultaneous feature selection and support vector machine optimization using the grasshopper optimization algorithm. *Cognitive Computation* pp. 1–18.
4. Aljarah, I., Faris, H., & Mirjalili, S. (2018). Optimizing connection weights in neural networks using the whale optimization algorithm. *Soft Computing, 22*(1), 1–15.
5. Aljarah, I., Faris, H., Mirjalili, S., & Al-Madi, N. (2018). Training radial basis function networks using biogeography-based optimizer. *Neural Computing and Applications, 29*(7), 529–553.
6. Aljarah, I., & Ludwig, S. A.: (2012). Parallel particle swarm optimization clustering algorithm based on mapreduce methodology. In *Proceedings of the Fourth World Congress on Nature and Biologically Inspired Computing (IEEE NaBIC12)*. IEEE Explore.
7. Aljarah, I., & Ludwig, S. A. (2013). A mapreduce based glowworm swarm optimization approach for multimodal functions. In *IEEE Symposium Series on Computational Intelligence, IEEE SSCI 2013*. IEEE Xplore.
8. Aljarah, I., & Ludwig, S. A.: A new clustering approach based on glowworm swarm optimization. In *Proceedings of 2013 IEEE Congress on Evolutionary Computation Conference (IEEE CEC13)*, Cancun, Mexico. IEEE Xplore.
9. Aljarah, I., & Ludwig, S. A. (2013). Towards a scalable intrusion detection system based on parallel pso clustering using mapreduce. In *Proceedings of Genetic and Evolutionary Computation Conference (ACM GECCO13) Amsterdam*, July 2013. ACM.
10. Aljarah, I., & Ludwig, S. A. (2016). A scalable mapreduce-enabled glowworm swarm optimization approach for high dimensional multimodal functions. *International Journal of Swarm Intelligence Research (IJSIR), 7*(1), 32–54.
11. Aljarah, I., Mafarja, M., Heidari, A. A., Faris, H., Zhang, Y., & Mirjalili, S. (2018). Asynchronous accelerating multi-leader salp chains for feature selection. *Applied Soft Computing, 71*, 964–979.

12. Aminisharifabad, M., Yang, Q., & Wu, X. (2018). A Penalized Autologistic regression with application for modeling the microstructure of dual-phase high strength steel. *Journal of Quality Technology*, in-press.
13. Asaithambi, S., & Rajappa, M. (2018). Swarm intelligence-based approach for optimal design of cmos differential amplifier and comparator circuit using a hybrid salp swarm algorithm. *Review of Scientific Instruments, 89*(5), 054702.
14. Barham, R., & Aljarah, I. (2017). Link prediction based on whale optimization algorithm. In *The International Conference on new Trends in Computing Sciences (ICTCS2017), Amman, Jordan.*
15. Barik, A. K., & Das, D. C. (2018). Active power management of isolated renewable microgrid generating power from rooftop solar arrays, sewage waters and solid urban wastes of a smart city using salp swarm algorithm. In *Technologies for Smart-City Energy Security and Power (ICSESP), 2018.* (pp. 1–6). IEEE.
16. Baygi, S. M. H., Karsaz, A., & Elahi, A. (2018). A hybrid optimal pid-fuzzy control design for seismic exited structural system against earthquake: A salp swarm algorithm. In *2018 6th Iranian Joint Congress on Fuzzy and Intelligent Systems (CFIS)* (pp. 220–225). IEEE.
17. Chitsaz, H., & Aminisharifabad, M. (2015). Exact learning of rna energy parameters from structure. *Journal of Computational Biology, 22*(6), 463–473.
18. Cho, J. H., Lee, D. J., & Chun, M. G. (2007). Parameter optimization of extreme learning machine using bacterial foraging algorithm. *Journal of Korean Institute of Intelligent Systems, 17*(6), 807–812.
19. Ding, S., Su, C., & Yu, J. (2011). An optimizing bp neural network algorithm based on genetic algorithm. *Artificial Intelligence Review, 36*(2), 153–162.
20. Ekinci, S., & Hekimoglu, B. (2018). Parameter optimization of power system stabilizer via salp swarm algorithm. In *2018 5th International Conference on Electrical and Electronic Engineering (ICEEE)* (pp. 143–147). IEEE.
21. El-Fergany, A. A. (2018). Extracting optimal parameters of pem fuel cells using salp swarm optimizer. *Renewable Energy, 119*, 641–648.
22. Eshtay, M., Faris, H., & Obeid, N. (2018). Improving extreme learning machine by competitive swarm optimization and its application for medical diagnosis problems. *Expert Systems with Applications, 104*, 134–152.
23. Eshtay, M., Faris, H., & Obeid, N. (2018). Metaheuristic-based extreme learning machines: a review of design formulations and applications. *International Journal of Machine Learning and Cybernetics* (pp. 1–19).
24. Faris, H., Mafarja, M., Heidari, A., Aljarah, I., Al-Zoubi, A., Mirjalili, S., et al. (2018). An efficient binary salp swarm algorithm with crossover scheme for feature selection problems. *Knowledge-Based Systems, 154*, 43–67.
25. Faris, H., Ala'M, A. Z., Heidari, A. A., Aljarah, I., Mafarja, M., Hassonah, M. A., et al. (2019). An intelligent system for spam detection and identification of the most relevant features based on evolutionary random weight networks. *Information Fusion, 48*, 67–83.
26. Faris, H., Aljarah, I., Al-Madi, N., & Mirjalili, S. (2016). Optimizing the learning process of feedforward neural networks using lightning search algorithm. *International Journal on Artificial Intelligence Tools, 25*(06), 1650033.
27. Faris, H., Aljarah, I., Al-Shboul, B. (2016). A hybrid approach based on particle swarm optimization and random forests for e-mail spam filtering. In *International Conference on Computational Collective Intelligence* (pp. 498–508). Springer, Cham.
28. Faris, H., Aljarah, I., & Mirjalili, S. (2016). Training feedforward neural networks using multiverse optimizer for binary classification problems. *Applied Intelligence, 45*(2), 322–332.
29. Faris, H., Aljarah, I., & Mirjalili, S. (2017). Evolving radial basis function networks using moth–flame optimizer. In *Handbook of Neural Computation* (pp. 537–550).
30. Faris, H., Aljarah, I., & Mirjalili, S. (2017). Improved monarch butterfly optimization for unconstrained global search and neural network training. *Applied Intelligence* pp. 1–20.
31. Faris, H., & Aljarah, I., et al. (2015). Optimizing feedforward neural networks using krill herd algorithm for e-mail spam detection. In *2015 IEEE Jordan Conference on Applied Electrical Engineering and Computing Technologies (AEECT)* (pp. 1–5). IEEE.

32. Faris, H., Hassonah, M. A., AlaM, A. Z., Mirjalili, S., & Aljarah, I. (2017). A multi-verse optimizer approach for feature selection and optimizing svm parameters based on a robust system architecture. *Neural Computing and Applications* pp. 1–15.
33. Ghatasheh, N., Faris, H., Aljarah, I., & Al-Sayyed, R. M. (2015). Optimizing software effort estimation models using firefly algorithm. *Journal of Software Engineering and Applications*, *8*(03), 133.
34. Gori, M., & Tesi, A. (1992). On the problem of local minima in backpropagation. *IEEE Transactions on Pattern Analysis & Machine Intelligence*, *1*, 76–86.
35. Gupta, J. N., & Sexton, R. S. (1999). Comparing backpropagation with a genetic algorithm for neural network training. *Omega*, *27*(6), 679–684.
36. Han, F., Yao, H. F., & Ling, Q. H. (2013). An improved evolutionary extreme learning machine based on particle swarm optimization. *Neurocomputing*, *116*, 87–93.
37. Heidari, A. A., Faris, H., Aljarah, I., & Mirjalili, S. (2018). An efficient hybrid multilayer perceptron neural network with grasshopper optimization. *Soft Computing*, 1–18.
38. Heidari, A. A., Kazemizade, O., & Hakimpour, F. (2017). A new hybrid yin-yang-pair swarm optimization algorithm for uncapacitated warehouse location problems. *ISPRS - International Archives of the Photogrammetry, Remote Sensing and Spatial Information Sciences XLII-4/W4* (pp. 373–379).
39. Heidari, A. A., Mirvahabi, S. S., & Homayouni, S. (2015). An effective hybrid support vector regression with chaos-embedded biogeography-based optimization strategy for prediction of earthquake-triggered slope deformations. *ISPRS - International Archives of the Photogrammetry, Remote Sensing and Spatial Information Sciences XL-1/W5* (pp. 301–305).
40. Heidari, A. A., & Abbaspour, R. A. (2018). Enhanced chaotic grey wolf optimizer for real-world optimization problems: A comparative study. In *Handbook of Research on Emergent Applications of Optimization Algorithms* (pp. 693–727). IGI Global.
41. Heidari, A. A., Abbaspour, R. A., & Jordehi, A. R. (2017). An efficient chaotic water cycle algorithm for optimization tasks. *Neural Computing and Applications*, *28*(1), 57–85.
42. Heidari, A. A., Abbaspour, R. A., & Jordehi, A. R. (2017). Gaussian bare-bones water cycle algorithm for optimal reactive power dispatch in electrical power systems. *Applied Soft Computing*, *57*, 657–671.
43. Heidari, A. A., & Delavar, M. R. (2016). A modified genetic algorithm for finding fuzzy shortest paths in uncertain networks. *ISPRS - International Archives of the Photogrammetry, Remote Sensing and Spatial Information Sciences XLI-B2* (pp. 299–304).
44. Heidari, A. A., & Pahlavani, P. (2017). An efficient modified grey wolf optimizer with lévy flight for optimization tasks. *Applied Soft Computing*, *60*, 115–134.
45. Huang, G. B., Zhu, Q. Y., & Siew, C. K. (2004). Extreme learning machine: A new learning scheme of feedforward neural networks. In *2004 IEEE International Joint Conference on Neural Networks, 2004. Proceedings*. vol. 2 (pp. 985–990). IEEE.
46. Huang, G. B., Zhu, Q. Y., & Siew, C. K. (2006). Extreme learning machine: Theory and applications. *Neurocomputing*, *70*(1), 489–501.
47. Hussien, A. G., Hassanien, A. E., & Houssein, E. H. (2017). Swarming behaviour of salps algorithm for predicting chemical compound activities. In *2017 Eighth International Conference on Intelligent Computing and Information Systems (ICICIS)* (pp. 315–320). IEEE.
48. Ismael, S., Aleem, S., Abdelaziz, A., & Zobaa, A. (2018). Practical considerations for optimal conductor reinforcement and hosting capacity enhancement in radial distribution systems. IEEE Access.
49. Lichman, M.: UCI machine learning repository (2013), http://archive.ics.uci.edu/ml.
50. Mafarja, M., & Abdullah, S. (2011). Modified great deluge for attribute reduction in rough set theory. In *2011 Eighth International Conference on Fuzzy Systems and Knowledge Discovery (FSKD)* (vol. 3, pp. 1464–1469). IEEE.
51. Mafarja, M., & Abdullah, S. (2013). Investigating memetic algorithm in solving rough set attribute reduction. *International Journal of Computer Applications in Technology*, *48*(3), 195–202.

52. Mafarja, M., & Abdullah, S. (2013). Record-to-record travel algorithm for attribute reduction in rough set theory. *Journal of Theoretical and Applied Information Technology, 49*(2), 507–513.
53. Mafarja, M., & Abdullah, S. (2014). Fuzzy modified great deluge algorithm for attribute reduction. In *Recent Advances on Soft Computing and Data Mining* (pp. 195–203). Springer, Cham.
54. Mafarja, M., & Abdullah, S. (2015). A fuzzy record-to-record travel algorithm for solving rough set attribute reduction. *International Journal of Systems Science, 46*(3), 503–512.
55. Mafarja, M., Aljarah, I., Heidari, A. A., Hammouri, A. I., Faris, H., AlaM, A. Z., et al. (2018). Evolutionary population dynamics and grasshopper optimization approaches for feature selection problems. *Knowledge-Based Systems, 145*, 25–45.
56. Mafarja, M., Aljarah, I., Heidari, A. A., Faris, H., Fournier-Viger, P., Li, X., et al. (2018). Binary dragonfly optimization for feature selection using time-varying transfer functions. *Knowledge-Based Systems, 161*, 185–204.
57. Mafarja, M., Jaber, I., Eleyan, D., Hammouri, A., & Mirjalili, S. (2017). Binary dragonfly algorithm for feature selection. In *2017 International Conference on New Trends in Computing Sciences (ICTCS)* (pp. 12–17).
58. Mafarja, M., & Mirjalili, S. (2017). Whale optimization approaches for wrapper feature selection. *Applied Soft Computing, 62*, 441–453.
59. Mirjalili, S., Gandomi, A. H., Mirjalili, S. Z., Saremi, S., Faris, H., & Mirjalili, S. M. (2017). Salp swarm algorithm: A bio-inspired optimizer for engineering design problems. *Advances in Engineering Software*.
60. Mirjalili, S., Mirjalili, S. M., & Lewis, A. (2014). Let a biogeography-based optimizer train your multi-layer perceptron. *Information Sciences, 269*, 188–209.
61. Mohapatra, P., Chakravarty, S., & Dash, P. K. (2015). An improved cuckoo search based extreme learning machine for medical data classification. *Swarm and Evolutionary Computation, 24*, 25–49.
62. Mohapatra, T. K., & Sahu, B. K. (2018). Design and implementation of ssa based fractional order pid controller for automatic generation control of a multi-area, multi-source interconnected power system. In *Technologies for Smart-City Energy Security and Power (ICSESP), 2018* (pp. 1–6). IEEE.
63. Reddy, Y. V. K., & Reddy, M. D. Solving economic load dispatch problem with multiple fuels using teaching learning based optimization and salp swarm algorithm. *Zeki Sistemler Teori ve Uygulamaları Dergisi, 1*(1), 5–15.
64. Sánchez-Monedero, J., Hervas-Martinez, C., Gutiérrez, P., Ruz, M. C., Moreno, M. R., & Cruz-Ramirez, M. (2010). Evaluating the performance of evolutionary extreme learning machines by a combination of sensitivity and accuracy measures. *Neural Network World, 20*(7), 899.
65. Sayed, G. I., Khoriba, G., & Haggag, M. H. (2018). A novel chaotic salp swarm algorithm for global optimization and feature selection. *Applied Intelligence* pp. 1–20.
66. Sexton, R. S., Dorsey, R. E., & Johnson, J. D. (1999). Optimization of neural networks: A comparative analysis of the genetic algorithm and simulated annealing. *European Journal of Operational Research, 114*(3), 589–601.
67. Shukri, S., Faris, H., Aljarah, I., Mirjalili, S., & Abraham, A. (2018). Evolutionary static and dynamic clustering algorithms based on multi-verse optimizer. *Engineering Applications of Artificial Intelligence, 72*, 54–66.
68. Wdaa, A. S. I. (2008). Differential evolution for neural networks learning enhancement. Ph.D. thesis, Universiti Teknologi Malaysia.
69. Xu, Y., & Shu, Y. (2006). Evolutionary extreme learning machine-based on particle swarm optimization. *Advances in Neural Networks-ISNN, 2006*, 644–652.
70. Yang, Z., Wen, X., Wang, Z. (2015). Qpso-elm: An evolutionary extreme learning machine based on quantum-behaved particle swarm optimization. In *2015 Seventh International Conference on Advanced Computational Intelligence (ICACI)* (pp. 69–72). IEEE.
71. Zhao, H., Huang, G., & Yan, N. (2018). Forecasting energy-related co2 emissions employing a novel ssa-lssvm model: Considering structural factors in china. *Energies, 11*(4), 781.

Sine Cosine Algorithm: Theory, Literature Review, and Application in Designing Bend Photonic Crystal Waveguides

Seyed Mohammad Mirjalili, Seyedeh Zahra Mirjalili, Shahrzad Saremi
and Seyedali Mirjalili

Abstract This chapter presented the Sine Cosine Algorithm (SCA), which is a recent meta-heuristics using mathematical equations to estimate the global optima of optimization problems. After discussing the mathematical model, a brief literature review is given covering the most recent improvements and applications of this algorithm. The performance of this algorithm is benchmarked on a wide range of test functions showing the flexibility of SCA in solving diverse problems with different characteristics. The chapter also considers finding an optimal design for a bend photonics crystal that shows the merits of this algorithm is solving challenging real-world problems.

1 Introduction

Meta-heuristics can be divided into two classes: population-based and individual-based. In the former class, an algorithm employs a population of solutions. This means that the algorithm starts with a set of random solutions. This set is then improved iteratively with several operators until the satisfaction of an end condition. In the

S. M. Mirjalili
Department of Electrical and Computer Engineering, Concordia University, Montreal, QC H3G 1M8, Canada
e-mail: mohammad.smm@gmail.com

S. Z. Mirjalili
School of Electrical Engineering and Computing, University of Newcastle, Callaghan, NSW 2308, Australia
e-mail: sz.mirjalili@gmail.com

S. Saremi · S. Mirjalili (✉)
Institute for Integrated and Intelligent Systems, Griffith University, Nathan, Brisbane, QLD 4111, Australia
e-mail: seyedali.mirjalili@griffithuni.edu.au

S. Saremi
e-mail: shahrzad.saremi@griffith.edu.au

© Springer Nature Switzerland AG 2020
S. Mirjalili et al. (eds.), *Nature-Inspired Optimizers*, Studies in Computational
Intelligence 811, https://doi.org/10.1007/978-3-030-12127-3_12

latter class, however, one solution is generated and improved until the end condition is met. Each of these classes have their own advantages and drawbacks.

Due to the use of multiple solutions, population-based meta-heuristics benefit from higher exploratory behaviours. In each iteration, more than one area of the search space is searched. Also, if a solution be trapped in a local solution, there are normally other solutions in better regions. In this case, the trapped solution is likely to be directed outside of the local solution. This leads to the second advantage, which is the information exchange between the solutions in the population. Due to the use of multiple solutions in population-based algorithms, solutions are aware of the quality of different regions of the search space, which puts them ahead of a single solution in an individual-based algorithm. Another advantage is less sensitivity to the initial population although uniform distribution across all the variables increases the exploration of the algorithm. The main drawback of population-based algorithms is the need to evaluation each solution in the population. The requires calling the objective function multiple times, which increase the run time of such algorithms. Another issue is the need to store the location and quality of some of the best solutions in each iteration to be used in the subsequent iteration. The last drawback worth mentioning here is the slow convergence speed of population-based algorithms.

On the other hand, individual-based algorithms require little space to store the single solution in each iteration. The number of function evaluations in such techniques is significantly less than that of population-based algorithms. Another advantage of such technique is the faster convergence speed although this might sometimes lead to premature convergence too. Despite these advantages, they normally suffer from local optima stagnation. This is because the exploratory behaviour of individual-based algorithm is less than population-based algorithms due to the use of less number of solutions.

The literature shows that population-based algorithms are used more than individual-based ones in a wide range of fields. This is because of the higher local optima avoidance of such algorithms. Population-based algorithms can be divided into multiple classes based on the source of inspirations: swarm-based, evolutionary, physics-based, event-based, and mathematics-based.

In swam-based algorithms, the collective behavior of a group of animals that leads to global intelligence without a centralized control unit is the main source of inspiration. This means people try to develop mathematical models of such swarming behavior and used them to proposed optimization algorithms. Some of the most well-regarded algorithms in this class are Particle Swarm Optimization [1], Ant Colony Optimization [2], and Bee Colony Optimization [3]. Some of the recent swarm-based algorithms are: Dragonfly Algorithm [4], Firefly Algorithm [5], Grey Wolf Optimizer [6], and Moth-Flame Algorithm [7] to name a few.

In evolutionary algorithms, as their names imply, the source of inspiration is from evolutionary phenomena in nature. Such algorithms simulate the way that organisms adapt to different environments. In fact, the main evolutionary operators in evolutionary algorithms are selection, recombination, and mutation [8, 9]. The way that these algorithms solve optimization problems is a bit different from swarm-based algorithms. Such methods maintain a constant rate of exploration and exploitation during optimization. However, swarm-based algorithms normally use adaptive mechanisms to tune exploration and exploitation. Some of the most well-regarded and recent evolutionary algorithms are: Genetic Algorithm [10], Differential Evolutionary [11], and Biogeography Optimization Algorithm [12].

In the physics-based class, a physical phenomenon is the main source of inspiration. Some inspirations are forces between solutions, interaction between particles, and particle movements. Some of the algorithms in this class are: Gravitational Search Algorithm [13], charged system search [14], Ray Optimization [15], and Multi-verse Optimizer [16]. In the event-based class, the inspiration can be from any event, phenomenon, or behaviour. The most popular algorithms in this class are: Imperialist Competition Algorithm (ICA) [17] and Teaching Learning Based Optimization (TLBO) [18].

The last class, mathematics-based, different mathematical models and equations are used to design optimization algorithms. One of the recent algorithms in this class is Since Cosine Algorithm (SCA) [19]. This algorithm employs a model based on trigonometric functions to solve optimization problems. This algorithm is discussed and analyzed in this chapter.

2 Sine Cosine Algorithm

The SCA algorithm [19] is population-based, in which a set of solutions is generated and improved using two main mathematical equations. The equations are as follows:

$$\overrightarrow{X_i^{t+1}} = \overrightarrow{X_i^t} + \overrightarrow{r_1} \sin(\overrightarrow{r_2}) \left| \overrightarrow{r_3} \overrightarrow{P_i^t} - \overrightarrow{X_i^t} \right| \tag{1}$$

$$\overrightarrow{X_i^{t+1}} = \overrightarrow{X_i^t} + \overrightarrow{r_1} \sin(\overrightarrow{r_2}) \left| \overrightarrow{r_3} \overrightarrow{P_i^t} - \overrightarrow{X_i^t} \right| \tag{2}$$

where $\overrightarrow{X_i^t}$ is a vector that represents the current solution including all the variables for the *i-th* solution in the *t-th* iteration, $\overrightarrow{X_i^t}$ is a vector that represents a destination including all the variables for the *i-th* solution in the *t-th* iteration, $\overrightarrow{r_1}$ is a random vector defining the magnitude of the range of *sin* and *cos*, $\overrightarrow{r_2}$ is a random variable to defined the domain of the sin or cos, and $\overrightarrow{r_3}$ indicates the magnitude of the destination contribution in defining the new position of the solution.

The above equations are designed to provide exploratory and exploitative behaviours for the SCA algorithm. A conceptual model of the search pattern when using these equations are given in Fig. 1.

Due to the use of *sin* and *cos* in the above equation this algorithm has been named Sine Cosine Algorithm. The SCA algorithm uses Eqs. 1 and 2 with 50% probability using the following equation:

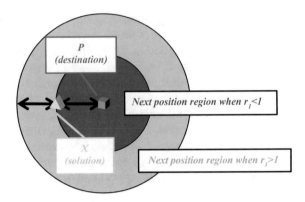

Fig. 1 In an $n - dimensional$ search space, the Eqs. 1 and 2 divide the search space into two spaces with respect a destination point (P): towards to outwards the destination. The main stochastic variable defining this movement direction is r_1

$$\overrightarrow{X_i^{t+1}} = \begin{cases} \overrightarrow{X_i^t} + \overrightarrow{r_1} \sin(\overrightarrow{r_2}) \left| \overrightarrow{r_3} \overrightarrow{P_i^t} - \overrightarrow{X_i^t} \right| & r_4 < 0.5 \\ \overrightarrow{X_i^t} + \overrightarrow{r_1} \cos(\overrightarrow{r_2}) \left| \overrightarrow{r_3} \overrightarrow{P_i^t} - \overrightarrow{X_i^t} \right| & r_4 \geq 0.5 \end{cases} \tag{3}$$

where r_4 is a random number in [0, 1].

The stochastic parameters of SCA play key roles in its performance. Each of them are defined for a purpose discussed as follows:

The parameter r_1 defines the magnitude of the range of sin and cos functions. It is the main mechanism to move the solution towards or outwards the destination. The impact of this parameter on the movement is shown in Fig. 2. This figure shows that when this parameter can map the range of sin and con in the interval of $[-2, 2]$.

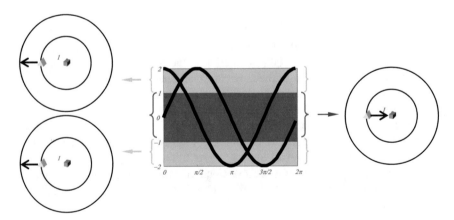

Fig. 2 The impact of r_1 on the movement of a solution with respect to a destination

When the trigonometric functions return a number in $[-2, 1]$ or $[1, 2]$, the solution steps away from the destination. When the functions return a value in $(-1, 1)$, the solution moves towards the destination.

Since exploration should be decreased in population-based algorithms to converge to a point at the end of optimization, the parameter r_1 linearly decreases in SCA as follows:

$$r_1 = a - t\frac{a}{T} \qquad (4)$$

where t shows the current iteration, a is a constant, and T shows the maximum number of iteration. The behaviour of this parameter when $a = 2$ and $T = 100$ is visualized in Fig. 3.

The impact of this linear parameter on the shape of sin and cos functions are shown in Fig. 4. This figure shows that both *sin* and *cos* functions do not return numbers greater than 1 or less than -1 when t is nearly 50th iteration. This is where solutions tend to merely move towards the destination.

The parameter r_1 shows the movement direction. The next parameter, r_2 indicates the step size of a solution towards or outwards the destination. The third parameter, r_3 defines the contribution level of the destination. When $r_3 > 1$ the destination significantly participate in the position updating process. This contribution decreases proportional to the iteration number. Finally, the last parameter, r_4, allows SCA to switch between the *sin* and *cos* function with an equal probability.

With the above equations, the SCA algorithm has been designed as follows to solve optimization problems:

The literature shows that the SCA algorithm has been widely used. The following section reviews the recent advances and applications and this technique.

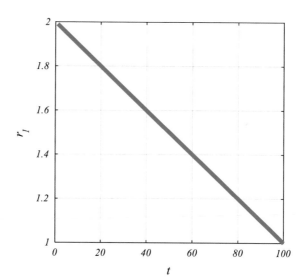

Fig. 3 The parameter r_1 is linearly decreased from 2 to 0

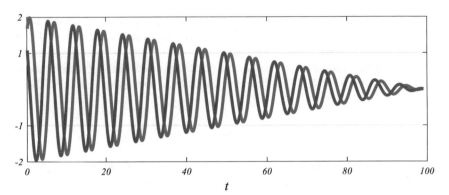

Fig. 4 The impact of this linear parameter on the shape of sin and cos functions

---

Algorithm 1 The SCA algorithm

Initialize a population of random solutions
while the end condition is not met **do**
   for Each solution **do**
      Evaluation the solution using the objective function
      if the objective value of the solution better than the destination (P) **then**
         Update P
      Update r_1, r_2, r_3, r_4
      Update the solution using Eq. 3

---

3 Literature Review of SCA

3.1 Variants

The SCA algorithm is able to solve problems with continuous variables. The problem should be unconstrained with one objective function to be solved by the original version of SCA. In the literature, there are several works to develop other variants of SCA as follows.

To solve multi-objective problems, Tawhid and Savsani proposed a multi-objective version of SCA using elitist non-dominated sorting and crowding distance approach in 2017 [20]. The former component allowed storing and improving non-dominated solutions obtained so far by SCA, and the latter component improves the diversity of non-dominated solutions.

To solve problems with binary and discrete variables, there are two works in the literature. In 2016, Hafez et al. [21] proposed a binary version of SCA using rounding to solve feature selection problems. The upper and lower bounds of each variables in SCA were set to 1 and 0 respectively. Before selecting each feature, these numbers were rounded to the nearest integer to select or discard a feature. In 2018, Reddy et al. [22] used a modified sigmoidal transformation function for binary mapping to solve binary problems using SCA.

As per the No Free Lunch theorem in optimization [23], there is no algorithm to solve all optimization problems. This holds for SCA as an optimization technique too, so there are many works in the literature trying to improve or tailor this algorithm to solve different classes of problems. The following paragraphs show the state-of-the-art improvements in SCA:

The SCA algorithm has been improved using Opposition-Based Learning (OBL). This mechanism allows an algorithm to use the opposite position of a solution in the search space either in the initialization phase or during optimization to increase exploration of an algorithm. SCA has been improved with both methods in [24, 25].

To improve exploratory and exploitative behaviours of SCA, different operators and mechanisms have been integrated into SCA as well. For instance, Levy flights were used in the position updating of SCA in 2017 [26] and 2018 [27]. Chaotic maps were used by Zou et al. so that the solutions in SCA shows more exploratory behaviours [28]. Another improvent has been the use of weighted position updating in SCA [29].

The SCA has been used in conjunction with machine learning techniques to solve a wide range of problems including clustering, classification, regression, and prediction as well. For instance, SCA was used to find optimal parameters for Support Vector Machine (SVM) in 2018 [30]. In the same year, SCA was hybridized with Extreme Learning Machine (ELM) for detection of pathological brain [31]. As a very population machine learning technique, Neural Networks are also trained, optimized by SCA in several works too [32–34].

The SCA algorithm has been hybridized with several other optimization algorithms too. The hybrid algorithms are:

- SCA + Particle Swarm Optimization [35, 36]
- SCA + Differential Evolution [37–40]
- SCA + Ant Lion Optimizer [41]
- SCA + Whale Optimization Algorithm (WOA) [42]
- SCA + Grey Wolf Optimizer (GWO) [43]
- SCA + water wave optimization algorithm [44]
- SCA + multi-orthogonal search strategy [45]
- SCA + Crow Search Algorithm [46]
- SCA + Teaching Learning Based Optimization [47]

3.2 Application

The SCA optimization algorithm has been widely used in different areas with or without modifications. The applications are summarized as follows:

- Re-entry trajectory optimization for space shuttle [48]
- Breast Cancer Classification [49]
- Power Distribution Network Reconfiguration [50]
- Temperature dependent optimal power flow [51]

- Pairwise Global Sequence Alignment [52]
- Tuning Controller Parameters for AGC of Multi-source Power Systems [53]
- Load frequency control of autonomous power system [54]
- Coordination of Heat Pumps, Electric Vehicles and AGC for Efficient LFC in a Smart Hybrid Power System [55]
- Economic and emission dispatch problems [56, 57]
- Optimization of CMOS analog circuits [58]
- Loss Reduction in Distribution System with Unified Power Quality Conditioner [59]
- Capacitive energy storage with optimized controller for frequency regulation in realistic multisource deregulated power system [60]
- Reduction of higher order continuous systems [61]
- Designing FO cascade controller in automatic generation control of multi-area thermal system incorporating dish-Stirling solar and geothermal power plants [62]
- SSSC damping Controller design in Power System [63]
- Selective harmonic elimination in five level inverter [64]
- Short-term hydrothermal scheduling [65]
- Optimal selection of conductors in Egyptian radial distribution systems [66]
- Data clustering [67]
- Loading margin stability improvement under contingency [68]
- Feature selection [69]
- Designing vehicle engine connecting rods [70]
- Designing single sensor-based MPPT of partially shaded PV system for battery charging [71]
- Handwritten Arabic manuscript image binarization [72]
- Thermal and Economical Optimization of a Shell and Tube Evaporator [73]
- Forecasting wind speed [74]
- Object tracking [39]

4 Results

In this section the SCA is first applied to several benchmark problems and compared to PSO and GA. The application of this technique in the field of photonics is investigated too.

4.1 Solving Benchmark Functions Using SCA

For solving the test functions, a total of 30 search agents are allowed to determine the global optimum over 500 iterations. The SCA algorithm is compared PSO and GA for verification of the results. Since the results of single run might be unreliable due to the stochastic nature of meta-heuristics, SCA, PSO, and GA are run 30 times

Table 1 Results of SCA on benchmark functions

F	SCA			PSO			GA		
	ave	Std	median	ave	Std	median	ave	std	median
F1	0.004965	0.007444	0.002375	2.429958	3.841589	0.725012	9035.277	1473.002	9563.743
F2	0.000301	0.000509	0.000118	5.209365	2.956499	4.823695	31.88797	3.492385	32.05133
F3	575.1372	484.6282	523.327	639.7214	480.5911	566.3505	15236.52	3182.203	14396.88
F4	9.991133	7.078092	8.301888	6.540452	2.075183	6.481008	47.31137	4.837602	48.24975
F5	92.09013	224.5418	19.74694	420.2764	617.9575	180.5272	7951559	2500670	7690166
F6	2.40103	0.309534	2.509621	3.862731	5.670347	0.54889	8247.615	1157.392	8165.055
F7	0.018688	0.022947	0.011776	0.147362	0.066498	0.130782	2.676552	0.898606	2.792091
F8	−3324.9	390.4823	−3381.72	−3053.54	638.4288	−2931.13	−4869.96	254.9391	−4807.78
F9	19.07492	24.20247	10.76504	43.33518	11.227	44.5402	125.9597	13.88292	129.1173
F10	7.5574	8.582238	4.227643	3.605528	1.544252	2.970032	16.42323	0.57628	16.39233
F11	0.292718	0.246861	0.256818	4.542635	2.084827	4.428388	80.94549	14.8369	76.76759
F12	0.711679	0.518935	0.501392	3.98657	3.524153	3.060955	5345707	3728346	5883151
F13	1.607044	0.434634	1.476889	5.465791	6.243191	2.881975	188868	7325210	188433
F14	125.9174	25.64531	123.6272	110	99.44289	100	141.0908	24.56276	130.5556
F15	116.7358	6.793422	115.919	171.5067	152.4691	115.9627	105.4678	12.25386	105.5092
F16	391.1075	74.17167	403.9587	243.4937	44.3344	232.2217	389.9803	81.71083	438.1829
F17	457.9372	49.95585	454.2577	474.5661	166.002	404.0431	484.6403	22.73743	490.0202
F18	138.6463	70.82028	99.261	196.7458	164.4387	202.7426	106.493	18.13203	103.8367
F19	526.6934	29.70283	535.8612	812.0401	167.2439	902.3394	545.5364	9.453189	545.9315

Table 2 P-values of the
Wilcoxon ranksum test over
all runs (p >= 0.05 rejects
the null hypothesis)

F	SCA	PSO	GA
F1	N/A	0.000183	0.000183
F2	N/A	0.000183	0.000183
F3	N/A	0.677585	0.000183
F4	0.307489	N/A	0.000183
F5	N/A	0.002202	0.000183
F6	N/A	0.140465	0.000183
F7	N/A	0.000246	0.000183
F8	0.000183	0.00044	N/A
F9	N/A	0.021134	0.000183
F10	0.850107	N/A	0.000183
F11	N/A	0.000183	0.000183
F12	N/A	0.004586	0.000183
F13	N/A	0.520523	0.000183
F14	0.384315	N/A	0.140167
F15	0.031209	0.472676	N/A
F16	0.000769	N/A	0.002202
F17	N/A	0.623176	0.053903
F18	0.909722	0.140465	N/A
F19	N/A	0.017257	0.017257

and statistical results (mean and standard deviation) are collected and reported in
Table 1. To decide about the significance of the results, a non-parametric statistical
test called Wilcoxon ranksum test is conducted as well. The p-values obtained from
this statistical test are reported in Table 2.

The results in Table 1 show that the SCA algorithm outperforms PSO and GA
on the majority of the test cases. Firstly, the SCA algorithm shows superior results
on six out of seven unimodal test functions. The p-values in Table 2 show that this
superiority is statistically significant on five out of six test functions. Due to the
characteristics of the unimodal test functions, these results strongly show that the
SCA algorithm has high exploitation and convergence. Secondly, Table 1 shows that
the SCA algorithm outperforms PSO and GA on the majority of the multi-modal test
functions (F9, F11, F12, and F13). The p-values in Table 2 also support the better
results of SCA statistically. These results prove that the SCA algorithm benefits from
high exploration and local optima avoidance. Finally, the results of the proposed
algorithm on the composite test functions in Tables 1 and 2 demonstrate the merits
of SCA in solving composite test functions with challenging search spaces.

4.2 Designing Bend Photonic Crystal Waveguides Using SCA

Photonic crystals (PhC) are periodic dielectric structures which show an interesting behaviour when interact with light. In a specific wavelength range, which is called photonic bandgap, light cannot transmit trough the structure. Therefore, when a light wave is transmitted toward the structure, it cannot pass and reflects back. This phenomenon is utilized to realize a waveguide. The problem which is investigate here is that when the waveguide is bended, some portion of light wave will be lost during the transmission. We utilize SCA to decrease the amount of lost of photonic crystal waveguide (PCW).

A typical form of sharp bend PCW is shown in Fig. 5 [75]. The rods are made from Si with refractive index of 3.6 and background is air. The lattice constant is 614 nm and the radius of typical rods is 114 nm. In the bend section, ellipse rods have been utilized to provide more flexibility to manipulate the transmitted light. Problem formulation of bend PCW designing is as follows.

$$f(\overrightarrow{x}) = -\text{Average of amplitude} \tag{5}$$

$$\overrightarrow{x} = \left[\frac{R_{a1}}{a}, \frac{R_{a2}}{a}, \frac{R_{a3}}{a}, \frac{R_{a4}}{a}, \frac{R_{a5}}{a}, \frac{R_{b1}}{a}, \frac{R_{b2}}{a}, \frac{R_{b3}}{a}, \frac{R_{b4}}{a}, \frac{R_{b5}}{a} \right] \tag{6}$$

$$\text{Subject to: } C_1 : 0 \leq \frac{R_{xx}}{a} \leq 0.5 \tag{7}$$

The SCA algorithm with 50 agents and a maximum iteration of 200 has been utilized to solve this problem. The average of the amplitude of the output spectral transmission is considered as the output of the objective function. The algorithm tries to minimize the $f(\overrightarrow{x})$ value. In Figs. 6 and 7 the convergence curve and the output spectral transmission performance of the final optimal bend PCW is shown.

Fig. 5 A typical form of sharp bend PCW. Nine rods considered to optimize by SCA

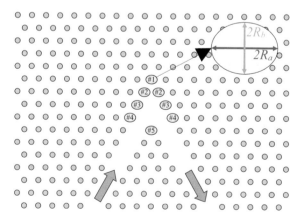

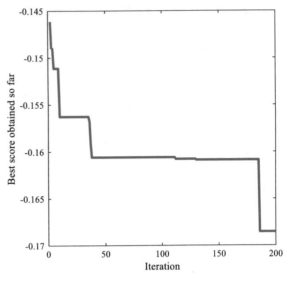

Fig. 6 Convergence curve of SCA for the bend PCW designing

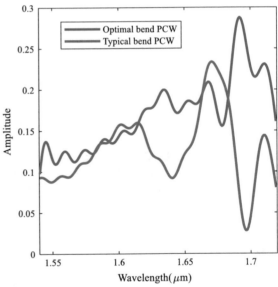

Fig. 7 Output spectral transmission performance of the final optimal bend PCW and typical form

The convergence curve shows that the SCA algorithm shows improved rate in the initial and final steps of optimization. The reason of significant improvements in the initial steps is the fact that the SCA algorithm works with a random population. The sine and cosine mechanisms as well as high values for the main controlling parameters cause solution to face abrupt changes and explore the search space. After nearly 50 iterations, the algorithm finds reasonable solutions so it tries to improve the accuracy. By the time that SCA reaches the final iterations, the controlling parameters

Table 3 Obtained optimal bend PCW structure with SCA (unit of R_{xx} is nanometer)

Label	R_{a1}	R_{a2}	R_{a3}	R_{a4}	R_{a5}	R_{b1}	R_{b2}	R_{b3}	R_{b4}	R_{b5}	$f(\vec{x})$
Optimized bend PCW	151	136	160	112	159	164	118	178	129	135	0.161
Typical bend PCW	114	114	114	114	114	114	114	114	114	114	0.122

Fig. 8 Physical geometry of the final optimal bend PCW design

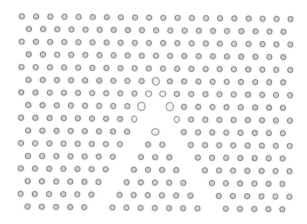

require the algorithm to change the solutions less abruptly, which leads to a local search and improvement in the accuracy of the solution found.

In Table 3, properties of the optimal design are shown. The comparison between the output of the typical bend PCW and optimal design of Fig. 7 shows that SCA finds a design that is nearly 30% better. The physical geometry of the final optimal design is also shown in Fig. 8. It can be seen that different ovals have found by SCA when tuning 10 parameters for this problem.

The results of the application show the merits of the SCA algorithm in solving challenging real-world problems. The SCA algorithm benefits from a small number of parameters to tune. The majority of its controlling parameters are tuned adaptively so the algorithm effectively balances exploration and exploitation when solving different problems.

5   Conclusion

This chapter presented the SCA algorithm as one of the recent meta-heuristics. The mathematical model of this algorithm was discussed and analyzed in detail. The chapter also considered benchmarking the performance of SCA in solving test functions. The results of this algorithm were compared with PSO and GA. It was observed that SCA outperforms both of these well-regarded algorithms significantly. The chapter also investigated the merits of SCA in solving a problem in the field of bend photonic crystal waveguides, which showed the merits of this algorithm in solving challenging problems.

References

1. Kennedy, J. (2011). Particle swarm optimization. In *Encyclopedia of machine learning* (pp. 760–766). Boston: Springer.
2. Dorigo, M., & Birattari, M. (2011). Ant colony optimization. In *Encyclopedia of machine learning* (pp. 36–39). Boston: Springer.
3. Karaboga, D., & Basturk, B. (2007). A powerful and efficient algorithm for numerical function optimization: Artificial bee colony (ABC) algorithm. *Journal of Global Optimization, 39*(3), 459–471.
4. Mirjalili, S. (2016). Dragonfly algorithm: A new meta-heuristic optimization technique for solving single-objective, discrete, and multi-objective problems. *Neural Computing and Applications, 27*(4), 1053–1073.
5. Yang, X. S. (2010). Firefly algorithm, stochastic test functions and design optimisation. *International Journal of Bio-Inspired Computation, 2*(2), 78–84.
6. Mirjalili, S., Mirjalili, S. M., & Lewis, A. (2014). Grey wolf optimizer. *Advances in Engineering Software, 69*, 46–61.
7. Mirjalili, S. (2015). Moth-flame optimization algorithm: A novel nature-inspired heuristic paradigm. *Knowledge-Based Systems, 89*, 228–249.
8. Chitsaz, H., & Aminisharifabad, M. (2015). Exact learning of rna energy parameters from structure. *Journal of Computational Biology, 22*(6), 463–473.
9. Aminisharifabad, M., Yang, Q., & Wu, X. (2018). A penalized autologistic regression with application for modeling the microstructure of dual-phase high strength steel. *Journal of Quality Technology* (in-press).
10. Holland, J. H. (1992). Genetic algorithms. *Scientific American, 267*(1), 66–73.
11. Neri, F., & Tirronen, V. (2010). Recent advances in differential evolution: A survey and experimental analysis. *Artificial Intelligence Review, 33*(1–2), 61–106.
12. Simon, D. (2008). Biogeography-based optimization. *IEEE Transactions on Evolutionary Computation, 12*(6), 702–713.
13. Rashedi, E., Nezamabadi-Pour, H., & Saryazdi, S. (2009). GSA: A gravitational search algorithm. *Information Sciences, 179*(13), 2232–2248.
14. Kaveh, A., & Talatahari, S. (2010). A novel heuristic optimization method: Charged system search. *Acta Mechanica, 213*(3–4), 267–289.
15. Kaveh, A., & Khayatazad, M. (2012). A new meta-heuristic method: Ray optimization. *Computers & Structures, 112*, 283–294.
16. Mirjalili, S., Mirjalili, S. M., & Hatamlou, A. (2016). Multi-verse optimizer: A nature-inspired algorithm for global optimization. *Neural Computing and Applications, 27*(2), 495–513.
17. Atashpaz-Gargari, E., & Lucas, C. (2007). Imperialist competitive algorithm: An algorithm for optimization inspired by imperialistic competition. In *IEEE Congress on Evolutionary Computation, 2007, CEC 2007* (pp. 4661–4667). IEEE.
18. Rao, R. V., Savsani, V. J., & Vakharia, D. P. (2011). Teachinglearning-based optimization: A novel method for constrained mechanical design optimization problems. *Computer-Aided Design, 43*(3), 303–315.
19. Mirjalili, S. (2016). SCA: A sine cosine algorithm for solving optimization problems. *Knowledge-Based Systems, 96*, 120–133.

20. Tawhid, M. A., & Savsani, V. (2017). Multi-objective sine-cosine algorithm (MO-SCA) for multi-objective engineering design problems. *Neural Computing and Applications*, 1–15.
21. Hafez, A. I., Zawbaa, H. M., Emary, E., & Hassanien, A. E. (2016). Sine cosine optimization algorithm for feature selection. In *International Symposium on Innovations in Intelligent Systems and Applications (INISTA), 2016* (pp. 1–5). IEEE.
22. Reddy, K. S., Panwar, L. K., Panigrahi, B. K., & Kumar, R. (2018). A new binary variant of sinecosine algorithm: Development and application to solve profit-based unit commitment problem. *Arabian Journal for Science and Engineering*, *43*(8), 4041–4056.
23. Wolpert, D. H., & Macready, W. G. (1997). No free lunch theorems for optimization. *IEEE Transactions on Evolutionary Computation*, *1*(1), 67–82.
24. Elaziz, M. A., Oliva, D., & Xiong, S. (2017). An improved opposition-based sine cosine algorithm for global optimization. *Expert Systems with Applications*, *90*, 484–500.
25. Bairathi, D., & Gopalani, D. (2017). Opposition-based sine cosine algorithm (OSCA) for training feed-forward neural networks. In *13th International Conference on Signal-Image Technology & Internet-Based Systems (SITIS), 2017* (pp. 438–444). IEEE.
26. Li, N., Li, G., & Deng, Z. (2017). An improved sine cosine algorithm based on levy flights. In *Ninth International Conference on Digital Image Processing (ICDIP 2017)* (Vol. 10420, p. 104204R). International Society for Optics and Photonics.
27. Qu, C., Zeng, Z., Dai, J., Yi, Z., & He, W. (2018). A modified sine-cosine algorithm based on neighborhood search and greedy levy mutation. *Computational Intelligence and Neuroscience*.
28. Zou, Q., Li, A., He, X., & Wang, X. (2018). Optimal operation of cascade hydropower stations based on chaos cultural sine cosine algorithm. In *IOP Conference Series: Materials Science and Engineering* (Vol. 366, No. 1, p. 012005). IOP Publishing.
29. Meshkat, M., & Parhizgar, M. (2017). A novel weighted update position mechanism to improve the performance of sine cosine algorithm. In *5th Iranian Joint Congress on Fuzzy and Intelligent Systems (CFIS), 2017* (pp. 166–171). IEEE.
30. Li, S., Fang, H., & Liu, X. (2018). Parameter optimization of support vector regression based on sine cosine algorithm. *Expert Systems with Applications*, *91*, 63–77.
31. Nayak, D. R., Dash, R., Majhi, B., & Wang, S. (2018). Combining extreme learning machine with modified sine cosine algorithm for detection of pathological brain. *Computers & Electrical Engineering*, *68*, 366–380.
32. Sahlol, A. T., Ewees, A. A., Hemdan, A. M., & Hassanien, A. E. (2016). Training feedforward neural networks using Sine-Cosine algorithm to improve the prediction of liver enzymes on fish farmed on nano-selenite. In *12th International Computer Engineering Conference (ICENCO), 2016* (pp. 35–40). IEEE.
33. Hamdan, S., Binkhatim, S., Jarndal, A., & Alsyouf, I. (2017). On the performance of artificial neural network with sine-cosine algorithm in forecasting electricity load demand. In *International Conference on Electrical and Computing Technologies and Applications (ICECTA), 2017* (pp. 1–5). IEEE.
34. Rahimi, H. (2019). Considering factors affecting the prediction of time series by improving Sine-Cosine algorithm for selecting the best samples in neural network multiple training model. In *Fundamental research in electrical engineering* (pp. 307–320). Singapore: Springer.
35. Chen, K., Zhou, F., Yin, L., Wang, S., Wang, Y., & Wan, F. (2018). A hybrid particle swarm optimizer with sine cosine acceleration coefficients. *Information Sciences*, *422*, 218–241.
36. Issa, M., Hassanien, A. E., Oliva, D., Helmi, A., Ziedan, I., & Alzohairy, A. (2018). ASCA-PSO: Adaptive sine cosine optimization algorithm integrated with particle swarm for pairwise local sequence alignment. *Expert Systems with Applications*, *99*, 56–70.
37. Bureerat, S., & Pholdee, N. (2017). Adaptive sine cosine algorithm integrated with differential evolution for structural damage detection. In *International Conference on Computational Science and Its Applications* (pp. 71–86). Cham: Springer.
38. Elaziz, M. E. A., Ewees, A. A., Oliva, D., Duan, P., & Xiong, S. (2017). A hybrid method of sine cosine algorithm and differential evolution for feature selection. In *International Conference on Neural Information Processing* (pp. 145–155). Cham: Springer.

39. Nenavath, H., & Jatoth, R. K. (2018). Hybridizing sine cosine algorithm with differential evolution for global optimization and object tracking. *Applied Soft Computing, 62,* 1019–1043.

40. Zhou, C., Chen, L., Chen, Z., Li, X., & Dai, G. (2017). A sine cosine mutation based differential evolution algorithm for solving node location problem. *International Journal of Wireless and Mobile Computing, 13*(3), 253–259.

41. Oliva, D., Hinojosa, S., Elaziz, M. A., & Ortega-Snchez, N. (2018). Context based image segmentation using antlion optimization and sine cosine algorithm. *Multimedia Tools and Applications,* 1–37.

42. Khalilpourazari, S., & Khalilpourazary, S. (2018). SCWOA: An efficient hybrid algorithm for parameter optimization of multi-pass milling process. *Journal of Industrial and Production Engineering, 35*(3), 135–147.

43. Singh, N., & Singh, S. B. (2017). A novel hybrid GWO-SCA approach for optimization problems. *Engineering Science and Technology, an International Journal, 20*(6), 1586–1601.

44. Zhang, J., Zhou, Y., & Luo, Q. (2018). An improved sine cosine water wave optimization algorithm for global optimization. *Journal of Intelligent & Fuzzy Systems, 34*(4), 2129–2141.

45. Rizk-Allah, R. M. (2018). Hybridizing sine cosine algorithm with multi-orthogonal search strategy for engineering design problems. *Journal of Computational Design and Engineering, 5*(2), 249–273.

46. Pasandideh, S. H. R., & Khalilpourazari, S. (2018). Sine cosine crow search algorithm: A powerful hybrid meta heuristic for global optimization. arXiv:1801.08485.

47. Nenavath, H., & Jatoth, R. K. Hybrid SCATLBO: A novel optimization algorithm for global optimization and visual tracking. *Neural Computing and Applications,* 1–30.

48. Banerjee, A., & Nabi, M. (2017). Re-entry trajectory optimization for space shuttle using Sine-Cosine algorithm. In *8th International Conference on Recent Advances in Space Technologies (RAST), 2017* (pp. 73–77). IEEE.

49. Majhi, S. K. (2018). An efficient feed foreword network model with sine cosine algorithm for breast cancer classification. *International Journal of System Dynamics Applications (IJSDA), 7*(2), 1–14.

50. Raut, U., & Mishra, S. Power distribution network reconfiguration using an improved sine cosine algorithm based meta-heuristic search.

51. Ghosh, A., & Mukherjee, V. (2017). Temperature dependent optimal power flow. In *2017 International Conference on Technological Advancements in Power and Energy (TAP Energy).* IEEE.

52. Issa, M., Hassanien, A. E., Helmi, A., Ziedan, I., & Alzohairy, A. (2018). Pairwise global sequence alignment using Sine-Cosine optimization algorithm. In *International Conference on Advanced Machine Learning Technologies and Applications* (pp. 102–111). Cham: Springer.

53. SeyedShenava, S., & Asefi, S. Tuning controller parameters for AGC of multi-source power system using SCA algorithm. *Delta, 2*(B2), B2.

54. Rajesh, K. S., & Dash, S. S. (2018). Load frequency control of autonomous power system using adaptive fuzzy based PID controller optimized on improved sine cosine algorithm. *Journal of Ambient Intelligence and Humanized Computing,* 1–13.

55. Khezri, R., Oshnoei, A., Tarafdar Hagh, M., & Muyeen, S. M. (2018). Coordination of heat pumps, electric vehicles and AGC for efficient LFC in a smart hybrid power system via SCA-based optimized FOPID controllers. *Energies, 11*(2), 420.

56. Mostafa, E., Abdel-Nasser, M., & Mahmoud, K. (2017). Performance evaluation of meta-heuristic optimization methods with mutation operators for combined economic and emission dispatch. In *2017 Nineteenth International Middle East Power Systems Conference (MEPCON)* (pp. 1004–1009). IEEE.

57. Singh, P. P., Bains, R., Singh, G., Kapila, N., & Kamboj, V. K. (2017). Comparative analysis on economic load dispatch problem optimization using moth flame optimization and sine cosine algorithms. No. 2, 65–75.

58. Majeed, M. M., & Rao, P. S. (2017). Optimization of CMOS analog circuits using sine cosine algorithm. In *8th International Conference on Computing, Communication and Networking Technologies (ICCCNT), 2017* (pp. 1–6). IEEE.

59. Ramanaiah, M. L., & Reddy, M. D. (2017). Sine cosine algorithm for loss reduction in distribution system with unified power quality conditioner. *i-Manager's Journal on Power Systems Engineering, 5*(3), 10.
60. Dhundhara, S., & Verma, Y. P. (2018). Capacitive energy storage with optimized controller for frequency regulation in realistic multisource deregulated power system. *Energy, 147,* 1108–1128.
61. Singh, V. P. (2017). Sine cosine algorithm based reduction of higher order continuous systems. In *2017 International Conference on Intelligent Sustainable Systems (ICISS)* (pp. 649–653). IEEE.
62. Tasnin, W., & Saikia, L. C. (2017). Maiden application of an sinecosine algorithm optimised FO cascade controller in automatic generation control of multi-area thermal system incorporating dish-Stirling solar and geothermal power plants. *IET Renewable Power Generation, 12*(5), 585–597.
63. Rout, B., PATI, B. B., & Panda, S. (2018). Modified SCA algorithm for SSSC damping controller design in power system. *ECTI Transactions on Electrical Engineering, Electronics, and Communications, 16*(1).
64. Sahu, N., & Londhe, N. D. (2017). Selective harmonic elimination in five level inverter using sine cosine algorithm. In *2017 IEEE International Conference on Power, Control, Signals and Instrumentation Engineering (ICPCSI)* (pp. 385–388). IEEE.
65. Das, S., Bhattacharya, A., & Chakraborty, A. K. (2017). Solution of short-term hydrothermal scheduling using sine cosine algorithm. *Soft Computing,* 1–19.
66. Ismael, S. M., Aleem, S. H. A., & Abdelaziz, A. Y. (2017). Optimal selection of conductors in Egyptian radial distribution systems using sine-cosine optimization algorithm. In *Nineteenth International Middle East Power Systems Conference (MEPCON), 2017* (pp. 103–107). IEEE.
67. Kumar, V., & Kumar, D. (2017). Data clustering using sine cosine algorithm: Data clustering using SCA. In *Handbook of research on machine learning innovations and trends* (pp. 715–726). IGI Global.
68. Mahdad, B., & Srairi, K. (2018). A new interactive sine cosine algorithm for loading margin stability improvement under contingency. *Electrical Engineering, 100*(2), 913–933.
69. Sindhu, R., Ngadiran, R., Yacob, Y. M., Zahri, N. A. H., & Hariharan, M. (2017). Sinecosine algorithm for feature selection with elitism strategy and new updating mechanism. *Neural Computing and Applications, 28*(10), 2947–2958.
70. Yldz, B. S., & Yldz, A. R. (2018). Comparison of grey wolf, whale, water cycle, ant lion and sine-cosine algorithms for the optimization of a vehicle engine connecting rod. *Materials Testing, 60*(3), 311–315.
71. Kumar, N., Hussain, I., Singh, B., & Panigrahi, B. K. (2017). Single sensor-based MPPT of partially shaded PV system for battery charging by using cauchy and gaussian sine cosine optimization. *IEEE Transactions on Energy Conversion, 32*(3), 983–992.
72. Elfattah, M. A., Abuelenin, S., Hassanien, A. E., & Pan, J. S. (2016). Handwritten arabic manuscript image binarization using sine cosine optimization algorithm. In *International Conference on Genetic and Evolutionary Computing* (pp. 273–280). Cham: Springer.
73. Turgut, O. E. (2017). Thermal and economical optimization of a shell and tube evaporator using hybrid backtracking search sine cosine algorithm. *Arabian Journal for Science and Engineering, 42*(5), 2105–2123.
74. Wang, J., Yang, W., Du, P., & Niu, T. (2018). A novel hybrid forecasting system of wind speed based on a newly developed multi-objective sine cosine algorithm. *Energy Conversion and Management, 163,* 134–150.
75. Jiang, L., Wu, H., Jia, W., & Li, X. (2013). Optimization of low-loss and wide-band sharp photonic crystal waveguide bends using the genetic algorithm. *Optik-International Journal for Light and Electron Optics, 124*(14), 1721–1725.

Whale Optimization Algorithm: Theory, Literature Review, and Application in Designing Photonic Crystal Filters

Seyedehzahra Mirjalili, Seyed Mohammad Mirjalili, Shahrzad Saremi and Seyedali Mirjalili

Abstract This chapter presents and analyzes the Whale Optimization Algorithm. The inspiration of this algorithm is first discussed in details, which is the bubble-net foraging behaviour of humpback whales in nature. The mathematical models of this algorithm is then discussed. Due to the large number of applications, a brief literature review of WOA is provided including recent works on the algorithms itself and its applications. The chapter also tests the performance of WOA on several test functions and a real case study in the field of photonic crystal filter. The qualitative and quantitative results show that merits of this algorithm for solving a wide range of challenging problems.

1   Introduction

Nature has been a major source of inspiration for researchers in the field of optimization. This has led to several nature- or bio-inspired algorithms. Such algorithms have mostly a similar framework. They start with a set of random solutions. This set is then improved using mechanisms inspired from nature. In Genetic Algorithms, for instance, each solution is considered as chromosomes in nature that is selected based

S. Mirjalili
School of Electrical Engineering and Computing, University of Newcastle, Callaghan, NSW 2308, Australia
e-mail: sz.mirjalili@gmail.com

S. M. Mirjalili
Department of Electrical and Computer Engineering, Concordia University, Montreal, QC H3G 1M8, Canada
e-mail: mohammad.smm@gmail.com

S. Saremi · S. Mirjalili (✉)
Institute for Integrated and Intelligent Systems, Griffith University, Nathan, Brisbane, QLD 4111, Australia
e-mail: seyedali.mirjalili@griffithuni.edu.au

S. Saremi
e-mail: shahrzad.saremi@griffith.edu.au

© Springer Nature Switzerland AG 2020
S. Mirjalili et al. (eds.), *Nature-Inspired Optimizers*, Studies in Computational Intelligence 811, https://doi.org/10.1007/978-3-030-12127-3_13

on its fitness value and undergoes crossover with other chromosomes and mutation. This simulates the process of natural selection, recombination, and mutation inspired from the theory of evolution. In Particle Swarm Optimization [1], each solution is considered as a vector that will be added to a velocity vector to simulate birds' flying method. In the Ant Colony Optimization [2], every solution is considered a path that an ant takes to reach certain goal. The method of finding the shortest path in a natural ant colony is then used to find an optimal solution for a given optimization problem.

Regardless of the differences between each of nature-inspired algorithms in the literature, one of the common features is the use of stochastic components. This means that a nature-inspired algorithm fluctuates solutions in a randomized manner to find the global optimum. Of course, the use of random components should be systematic to perform better than a complete random search. One of the ways to increase the chance of finding the optimal or near optimal solutions using a nature-inspired algorithm that leads to a systematic stochastic search is to change the magnitude or rate of changes during the optimization process. Most of nature-inspired algorithms divide the search process into the phases of exploration and exploitation [3].

In the exploration phase, which is also called diversification, an algorithm abruptly changes the solutions. This means using random components is at its maximum. The main reason is to explore the search space. The search space of a problem is n-dimensional where n is the number of variables. Depending on the nature of those variables, the search space might be definite (in case discrete variables) or indefinite (in case of continuous variables). In either cases, an algorithm needs to search the most promising regions to find the global optimum. This can be achieving by randomly changing the variables.

Exploration should be followed by exploitation, which is often called intensification. After finding the promising regions of a search space, a nature-inspired algorithms should locally search them to improve the accuracy. This can be achieved by reducing the magnitude or rate of random changes in the solutions. Finding a good balance between exploration and exploitation is challenging for a nature-inspired algorithm. For the exploitation, any sort of local search can be used too, but the main challenge is when to start the local search. To achieved a balance, some algorithms use adaptive numbers and mechanism.

The literature shows that bio-inspired algorithms are reliable alternatives to the conventional optimization algorithms (e.g. gradient-based) despite the fact that they belong to the family of stochastic algorithms [4, 5]. This is mainly due to the better local solution avoidance of these algorithms and problem independency as compared to conventional optimization and search algorithms.

Optimization problems can be divided into two classes based on their shape of search spaces: linear versus non-linear. In the first class, there is no local solution, so a gradient-based technique is the best algorithm and significantly outperforms stochastic optimization algorithms. In non-linear problems, however, there might be a large number of local solutions that should be avoided to find the global optimum. Gradient-based algorithms tend to get stuck in the local solutions since they move towards the negative of gradient. In this case, the stochastic components of nature-inspired algorithms allow them to better avoid those local solutions. They do not need gradient information of a problem, so the range of their applications is substantially wider than that of conventional algorithms.

One of the recent nature-inspired algorithm is the Whale Optimization Algorithm (WOA) proposed in 2016 [6]. This algorithm mimics the unique foraging behaviour of humpback whales in nature. This chapter covers the details of inspiration, mathematical models, and the main operators of this algorithm. The chapter then analyzes the performance of this algorithm on several mathematical benchmark functions. The application of WOA in the area of photonic crystal filter design is investigate too.

2 Whale Optimization Algorithm

This WOA algorithm is inspired from the bubble-net foraging behaviour of humpback whales in nature. Humpback whales are not fast enough to chase and consume a school of fish in oceans. This required their evolutionary process to figure out techniques to trap preys instead. Such creatures use bubbles to trap fish. In this mechanism, one or multiple whales swarm in a spiral-shaped path around a school of fish while making bubbles. As can be seen in Fig. 1, this method directs the school towards the surface. The spiral movement radius becomes smaller and smaller towards the surface of water. When the school is very close to the surface, humpback whales attack them. This mechanism is simulated in the WOA algorithm as the main operator.

The main mechanisms implemented in the WOA algorithm are as follows:

- Encircling the prey
- Bubble-net attacking
- Search for the prey

Fig. 1 Bubble-net foraging behavior of humpback whales in nature

In WOA, whales swim in an n-dimensional search space where n indicates the number of variables. In other words, a whale is a candidate solution made of n variables. The first step to find the global solution is to update the position of each solution in the search space. This is done using the following equations:

$$\overrightarrow{X(t+1)} = \overrightarrow{X^*}(t) - \overrightarrow{A} \cdot \left| \overrightarrow{C} \cdot \overrightarrow{X^*(t)} - \overrightarrow{X}(t) \right| \qquad (1)$$

where $\overrightarrow{X(t)}$ shows is the solution vector in the t-th iteration, $\overrightarrow{X^*(t)}$ is the possible position of the prey in the t-th iteration, $\cdot$ is a pairwise multiplication between two vectors, and $\overrightarrow{A}$, $\overrightarrow{C}$ are parameters vectors calculated for each dimension and updated in each iteration as follows:

$$\overrightarrow{A} = 2\overrightarrow{a} \cdot \overrightarrow{r_1} - \overrightarrow{a} \qquad (2)$$

$$\overrightarrow{C} = 2 \cdot \overrightarrow{r_2} \qquad (3)$$

where $\overrightarrow{a}$ linearly decreases from 2 to 0 during the optimization process generated for each dimension, and $\overrightarrow{r_1}$, $\overrightarrow{r_2}$ are random vectors in [0, 1] for each dimension.

The above equation allows whales (solutions) to relocate to any areas around a given prey $(\overrightarrow{X^*(t)})$. The random components allow movement in a hyper-rectangle around a prey as shown in Fig. 2.

The spiral movement of whales in the WOA algorithm is done using the following equation:

$$\overrightarrow{X}(t+1) = \overrightarrow{D'} e^{bt} cos(2a\pi t) + \overrightarrow{X^*(t)} \qquad (4)$$

The shape of this spiral and its impact on the movement of whales are shown in Fig. 3.

The process of encircling the prey and spiral movement is done in turn in the WOA algorithm. This is simulated using the following equation.

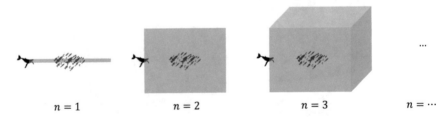

$n = 1$ $\qquad$ $n = 2$ $\qquad$ $n = 3$ $\qquad$ $n = \cdots$

Fig. 2 The position updating process in n-dimensional spaces using Eq. 1

Fig. 3 The spiral movement
of whales in the WOA
algorithm

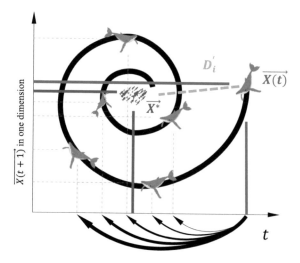

$$
\overrightarrow{X(t+1)} = \begin{cases} \overrightarrow{X^*(t+1)} - \overrightarrow{A} \cdot \overrightarrow{D} & r_3 < 0.5 \\ \overrightarrow{D'}ebt\cos(2a\pi t) + \overrightarrow{X^*(t)} & r_3 > 0.5 \end{cases} \tag{5}
$$

where r_3 is a random number in $[0.1]$.

This equation gives 50% chance to each of the position updating mechanism for
each value in each iteration and for each dimension. The WOA algorithm is shown
in Algorithm 1.

Algorithm 1 The WOA algorithm

Initialize a set of random solutions
Initialize $\overrightarrow{a}$, $\overrightarrow{A}$, and $\overrightarrow{C}$
while t **do**he end condition is not met
    Calculate the objective value of the population
    Find the best solution and updated $\overrightarrow{X^*}$
    if $-1 < \overrightarrow{A} < 1$ **then**
        Update the position using Eq. 5
    else
        Use a randomly selected solution instead of $\overrightarrow{X^*}$
    end if
    Update $\overrightarrow{a}$, $\overrightarrow{A}$, and $\overrightarrow{C}$
end while

Note that in the WOA algorithm, $\overrightarrow{X^*}$ is chosen to move around randomly selected
whales instead of the best one to improve exploration.

3 Literature Review of WOA

This section provides a brief review of the most recent developments of the WOA algorithm and its applications.

3.1 Variants

The WOA algorithm requires solutions to move in a continuous search space. The solutions are evaluated with an objective function and compared using relational operators to find the best one. This means that WOA is suitable for solving unconstrained single-objective optimization problems with continuous variables. The original version of this algorithm have been widely used and improved.

The parameters of WOA have been tuned to improve its performance. The literature shows that a popular technique has been the use of chaotic maps [7–10] to tune the parameters of WOA. In such methods, chaotic maps are used to generate numbers for the random components of algorithm where there is a need to generated random numbers in [0, 1]. This allows more chaotic and randomness behaviours that leads to higher exploration for an optimization algorithm.

Despite the fact that there is one adaptive parameter in WOA to balance exploration and exploitation, there are some works in the literate trying to use other adaptive mechanisms [11]. For instance, an inertia weight is used for WOA in [12]. Another example is the use of adaptive random walks. Emary et al. [13] proposed this improved WOA and showed that the performance of WOA can be significantly improved. Most of these works argued that integrating random components increases exploration of WOA.

Opposition-based learning WOA was proposed in [14] for the same reasons. The authors considered the 'opposite' position of whales as an alternative methods of position updating. It was shown that the performance of WOA is increased with the opposition-based learning. In the literature, levy flight random flights have been used to updated the position of whales [15, 16] as well. Since levy flight is a local search, it was shown in these two studies that the exploitation and convergence speed of WOA can be increased.

The above-mentioned algorithms are able to solve continuous problems with no constraints and one objective. To solve other types of problems using WOA, there have been several attempts in the literature. For instance, a penalty function was used to handle constrains when using WOA in [17]. A penalty function is normally algorithm-independent since it is integrated to the objective function. The binary version of WOA was proposed in [18–21]. The authors of these works use crossover or transfer function to move solutions of WOA in a binary search space.

To solve multi-objective problems with the WOA algorithm, there have been five attempts in the literature [22–26]. The main mechanisms in these works are aggregation of objectives, non-dominated sorting, and archives.

The field of meta-heuristics is full of homogeneous hybridization methods. In such works, authors try to combine multiple meta-heuristics to benefit from the advantages of the algorithms when solving problems. The hybridization might be also heterogeneous, meaning that one of the algorithms is of a different type. For example, a machine learning algorithm can be hybridized with a meta-heuristic to solve optimization or classification problems. Some of the popular hybrids using WOA are as follows.

Homogeneous hybrids:

- WOA +PSO [27]
- WOA + DE [28]
- WOA + GWO [29–31]
- WOA + BO [32]
- WOA + SCA [33]

Heterogeneous hybrids:

- WOA + SVM [34]
- WOA + MLP [35]
- WOA + NN [36, 37]

3.2 Applications

The WOA algorithm has been widely used in different fields. Due to the large number of publications, some of them are listed in this subsection as follows:

- Energy [38–43]
- Computer networks [44–50]
- Cloud computing [51–53]
- Image processing and machine vision [54–57]
- Electronics and electrical engineering [58–61]
- Antenna design [62, 63]
- Clustering [64–66]
- Prediction [67–69]
- Classification [70–72]
- Management [73, 74]
- Feature selection [75–80]
- Aerospace [81]
- Medicine [82–85]
- Robotics [86, 87]
- Structural optimization [88–90]
- Operation Research [91–97]

4 Results

In this section the WOA algorithm is applied to a number of test functions and a real-world problem.

4.1 Results of WOA on Benchmark Functions

The results of WOA on test functions are presented in Figs. 4, 5, and 6. Figure 4 shows the performance of WOA on a set of linear problems (unimodal). The search history of whales shows that the algorithm benefits from a good coverage of search

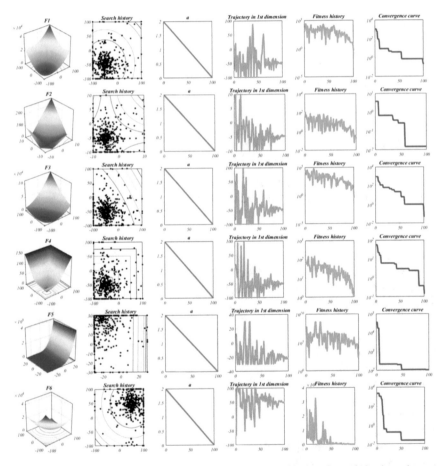

Fig. 4 Results of WOA when solving unimodal test functions: history of sampled points, changes in the parameter a, trajectory of one variable in the first solution, average objective of all solutions, and convergence curve

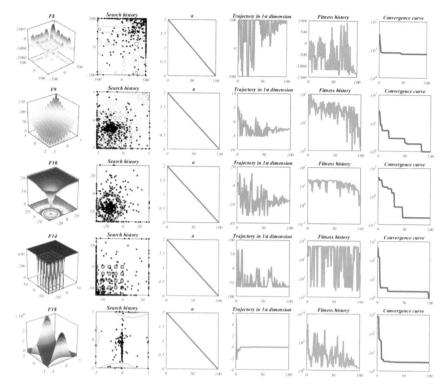

Fig. 5 Results of WOA when solving multimodal test functions: history of sampled points, changes in the parameter a, trajectory of one variable in the first solution, average objective of all solutions, and convergence curve

space. The search history also shows that the sampled points tend to more scatter around the global optimum. The fourth column in Fig. 4 shows the fluctuation in the variables of solutions. It can be seen that abrupt changes face away proportional to the number of iterations. This is due to the adaptive values for the parameter a. The third column in Fig. 4 shows that this parameter is a function of the iteration and reduces over time. This reduces the magnitude of movements for the solutions. The fifth column in Fig. 4 shows the average objective value of all solutions. It can be seen that this average tends to be decrease, showing the success of WOA in improving the quality of the first initial population. Finally, the convergence curves in the last column shows significant improvement in the best solution obtained so far too. The curves tend to provide consistent rate, which shows that WOA benefits from a high accuracy of the results obtained.

The results of WOA on multi-modal test functions are given in Fig. 5. Most of the subplots are consistent with those in Fig. 4. The most interesting pattern is rapid changes in both trajectory of variable and average fitness functions of all solutions. This is due to the large number of local solutions and non-linear search spaces of

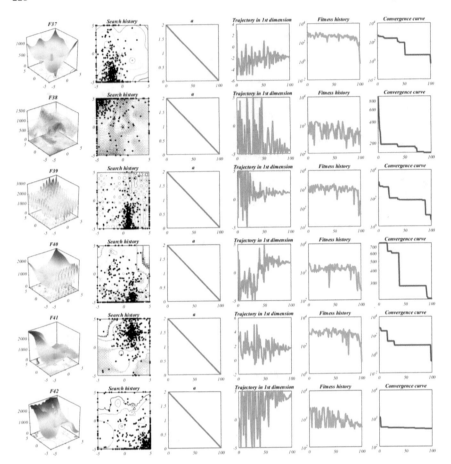

Fig. 6 Results of WOA when solving composite test functions: history of sampled points, changes in the parameter a, trajectory of one variable in the first solution, average objective of all solutions, and convergence curve

these problems. The number of times that the best solution obtained so far gets updated is less than that in the unimodal test functions. Therefore, the solutions tend to explore more to find better solutions. Despite this fact, the overall improvements in both average fitness and best fitness value of the best solution are evident. The convergence curves also show less improvement than those obtained for the unimodal test functions. This is again due to the periods of times that WOA is searching to find a better and more promising solution in the search space.

The results of WOA on composite test functions are give in Fig. 6. The results are very similar to those on multi-modal test functions. However, the convergence curves tend to stay constant for a large number of iterations. Composite test functions are non-uniform, non-symmetrical, and highly unbiased. This makes them very difficult to solve. An algorithm needs to maintain a high level of randomness to constantly avoid the large number of local solutions in such problems. Higher exploratory behaviour can be seen in the trajectories as well.

Overall, the results on the benchmark functions show the search patterns of the WOA algorithm. As per the results, it can be stated that the WOA algorithm benefits from a good balance of exploration and exploitation to solve linear and nonlinear problems. However, this does not mean that it can solve all optimization problems according to the No Free Lunch theorem. In the next subsection, this algorithm is applied to a real-world problem in the area of photonic crystal filter design to showcase its performance and flexibility in solving challenging, real-world problems too.

4.2 Results of WOA When Designing Photonic Crystal Filters

Devices made of Photonic Crystals (PhC) have become popular lately, which is mainly due to the bandgap phenomenon of PhC structures and defection creation that lead to manipulate the light for different applications. However, modeling the behavior of light propagation in such devices is challenging. Due to the lack of analytical model for designing such devices, designers normally use trial and error to design them [98].

There are some works in the literature that attempt to employ optimization algorithms to design devices. For example, Genetic Algorithms used in [99] to maximize the Quality factor (Q) of PhC cavity. Another algorithm used in this area is Grey Wolf Optimizer [100]. This section the WOA algorithm is employed to design a filter made of PhC structure.

The proposed structure of PhC filter to be optimized by WOA is visualized in Fig. 7. This figure shows that there are some holes on a silicon slab to create a waveguide and a cavity section. As one of the main parts, the cavity is made of seven holes. Light enters from the right hand side and the structure provide filtering behavior by the cavity section. After exiting the light form the right hand side, the spectral transmission can be measured.

The main role of the WOA algorithm will be to find an optimal radius for the holes in the cavity. The radius of the other holes considered constant during the optimization because the their duty is to guide the light in the device.

The thickness of the silicon is 400 nm. The radius of the typical holes and lattice constant of the PhC lattice are considered as 162 and 433 nm. For more information regarding the structure and how it works, interested readers are referred to [101–103]. To calculate the output performance of the filter, a 2D FDTD simulation is used for TE-polarization. An example can be seen in Fig. 8. In order to design a high performance filter, there merit factors must be considered as shown in Fig. 8 and explained as follows [101–103].

Fig. 7 Proposed PhC filter.
Seven holes are used to form
the super defect region

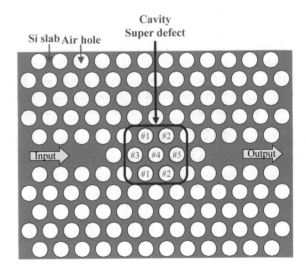

Fig. 8 Output spectral
transmission performance of
a sample case of PhC filter

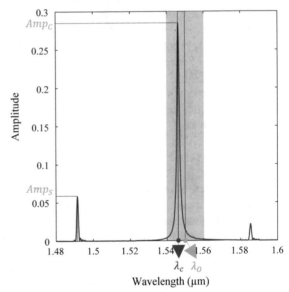

Wavelength (μm)

The objectives involved in this problems are as follows:

- Amp_c: Maximum amplitude of the output in the main band
- AMP_s: Maximum amplitude of the output in the sidebands
- *Deviation*: The deviation of the central wavelength of the output peak (λ_c) to the defined central wavelength of the channel (λ_O) ($|\lambda_c - \lambda_O|$)

Amp_C should be maximized while the Amp_S and Deviation must be minimized. To solve this problem with single objective WOA, an aggregation of these merit factor will be considered as the objective function.

The problem of designing the PhC filter is formulated as follows:

$$\text{Minimize:} f(\vec{x}) = \frac{-Amp_c}{Amp_s + deviation} \tag{6}$$

$$\vec{x} = \left[\frac{R_1}{a}, \frac{R_2}{a}, \frac{R_3}{a}, \frac{R_4}{a}, \frac{R_5}{a} \right] \tag{7}$$

$$\text{Subject to:} C_1 : 0 \le \frac{R_1}{a}, \frac{R_2}{a}, \frac{R_3}{a}, \frac{R_4}{a}, \frac{R_5}{a} \le 0.5 \tag{8}$$

$$C_2 : Amp_s < Amp_c \tag{9}$$

The problem formulation shows that there are five parameters to optimize. The parameters define the radii of cavity holes in the slab as shown in Fig. 7. To solve this problem, 50 whales and 200 iterations are used. The optimal values found for the parameters are shown in Table 1.

Table 1 Optimal values found for the parameters and the objective value using 50 whales and 200 iterations

R_1	R_2	R_3	R_4	R_5	O
35	100	160	23	2	−9.91

Fig. 9 Convergence curve of WOA for the PhC filter designing

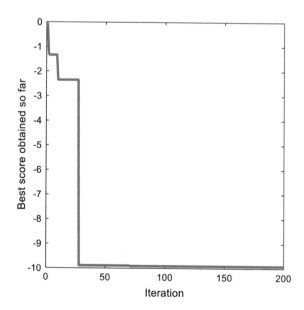

The convergence curve and the output spectral transmission performance of the final optimal PhC filter and its physical geometry are also shown in Figs. 9, 10, and 11. The convergence curve shows that the WOA algorithm quickly converges towards a good solution in the first 30 iteration. There are significant drops in the objective values in this curve. After this, the convergence curve is very slow. The WOA algorithm was run multiple times and the same behavior was observed. So, it can be said that this is due to the nature of this problem. The optimal design found at the end of iteration 200 is desirable as the output spectral transmission performance shown in Fig. 10.

That optimal shape found by WOA for the slab in Fig. 11 shows that the holes have different radii. The biggest one is on the left hand side. On the other hand, the

Fig. 10 Output spectral transmission performance of the final optimal PhC filter designs

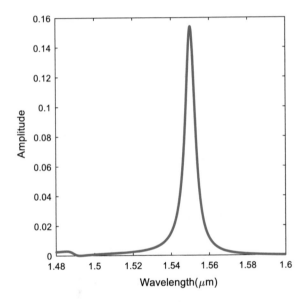

Fig. 11 Physical geometry of the final optimal PhC filter design

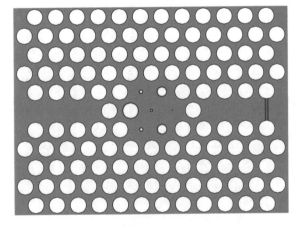

smallest hole is the one on the right hand side. Such diverse shapes for the holes are obtained by WOA to minimize the objective function. The results of this subsection showed that the WOA algorithm can be beneficial when applying the challenging real-world problems too.

5 Conclusion

This work presented the Whale Optimization Algorithm as one of the most recent Swarm Intelligence techniques. The inspiration of this algorithm was discussed first. Then, the mathematical model and algorithm were discussed in detail. To analyze the performance of this algorithm, several experiments were conducted on benchmark functions. The results showed that the WOA algorithm benefits from a good balance of exploration and exploitation. This assisted this algorithm to avoid local solutions in a search space and find a very accurate estimation of the global optima for the test functions. The chapter also applied WOA to design a photonic crystal filter as a real-world case study. It was observed that the WOA algorithm find an optimal design despite the challenging nature of the case study.

References

1. Kennedy, J. (2011). Particle swarm optimization. In *Encyclopedia of machine learning* (pp. 760–766). Boston, MA: Springer.
2. Dorigo, M., & Birattari, M. (2011). Ant colony optimization. In *Encyclopedia of machine learning* (pp. 36–39). Boston, MA: Springer.
3. Tan, K. C., Chiam, S. C., Mamun, A. A., & Goh, C. K. (2009). Balancing exploration and exploitation with adaptive variation for evolutionary multi-objective optimization. *European Journal of Operational Research, 197*(2), 701–713.
4. Chitsaz, H., & Aminisharifabad, M. (2015). Exact learning of rna energy parameters from structure. *Journal of Computational Biology, 22*(6), 463–473.
5. Aminisharifabad, M., Yang, Q. & Wu, X. (2018). A penalized autologistic regression with application for modeling the microstructure of dual-phase high strength steel. *Journal of Quality Technology* (in-press).
6. Mirjalili, S., & Lewis, A. (2016). The whale optimization algorithm. *Advances in Engineering Software, 95*, 51–67.
7. Oliva, D., El Aziz, M. A., & Hassanien, A. E. (2017). Parameter estimation of photovoltaic cells using an improved chaotic whale optimization algorithm. *Applied Energy, 200*, 141–154.
8. Sun, W. Z., & Wang, J. S. (2017). Elman neural network soft-sensor model of conversion velocity in polymerization process optimized by chaos whale optimization algorithm. *IEEE Access, 5*, 13062–13076.
9. Kaur, G., & Arora, S. (2018). Chaotic whale optimization algorithm. *Journal of Computational Design and Engineering*.
10. Prasad, D., Mukherjee, A., & Mukherjee, V. (2017). Transient stability constrained optimal power flow using chaotic whale optimization algorithm. In *Handbook of neural computation* (pp. 311–332).

11. Trivedi, I. N., Pradeep, J., Narottam, J., Arvind, K., & Dilip, L. (2016). Novel adaptive whale optimization algorithm for global optimization. *Indian Journal of Science and Technology*, *9*(38).
12. Hu, H., Bai, Y., & Xu, T. (2016). A whale optimization algorithm with inertia weight. *WSEAS Transactions on Computers*, *15*, 319–326.
13. Emary, E., Zawbaa, H. M., & Salam, M. A. (2017). A proposed whale search algorithm with adaptive random walk. In *2017 13th IEEE International Conference on Intelligent Computer Communication and Processing (ICCP)* (pp. 171–177). IEEE.
14. Elaziz, M. A., & Oliva, D. (2018). Parameter estimation of solar cells diode models by an improved opposition-based whale optimization algorithm. *Energy Conversion and Management*, *171*, 1843–1859.
15. Ling, Y., Zhou, Y., & Luo, Q. (2017). Lévy flight trajectory-based whale optimization algorithm for global optimization. *IEEE Access*, *5*(99), 6168–6186.
16. Abdel-Basset, M., Abdle-Fatah, L., & Sangaiah, A. K. (2018). An improved Lévy based whale optimization algorithm for bandwidth-efficient virtual machine placement in cloud computing environment. *Cluster Computing*, 1–16.
17. Sauber, A. M., Nasef, M. M., Houssein, E. H., & Hassanien, A. E. (2018). Parallel whale optimization algorithm for solving constrained and unconstrained optimization problems. arXiv:1807.09217.
18. Eid, H. F. (2018). Binary whale optimisation: an effective swarm algorithm for feature selection. *International Journal of Metaheuristics*, *7*(1), 67–79.
19. Hussien, A. G., Houssein, E. H., & Hassanien, A. E. (2017, December). A binary whale optimization algorithm with hyperbolic tangent fitness function for feature selection. In *2017 Eighth International Conference on Intelligent Computing and Information Systems (ICICIS)* (pp. 166–172). IEEE.
20. Reddy K, S., Panwar, L., Panigrahi, B. K., & Kumar, R. (2018). Binary whale optimization algorithm: A new metaheuristic approach for profit-based unit commitment problems in competitive electricity markets. *Engineering Optimization*, 1–21.
21. Hussien, A. G., Hassanien, A. E., Houssein, E. H., Bhattacharyya, S., & Amin, M. (2019). S-shaped binary whale optimization algorithm for feature selection. In *Recent trends in signal and image processing* (pp. 79–87). Singapore: Springer.
22. Wang, J., Du, P., Niu, T., & Yang, W. (2017). A novel hybrid system based on a new proposed algorithm-multi-objective whale optimization algorithm for wind speed forecasting. *Applied Energy*, *208*, 344–360.
23. El Aziz, M. A., Ewees, A. A., & Hassanien, A. E. (2018). Multi-objective whale optimization algorithm for content-based image retrieval. *Multimedia Tools and Applications*, 1–38.
24. El Aziz, M. A., Ewees, A. A., Hassanien, A. E., Mudhsh, M., & Xiong, S. (2018). Multi-objective whale optimization algorithm for multilevel thresholding segmentation. In *Advances in soft computing and machine learning in image processing* (pp. 23–39). Cham: Springer.
25. Jangir, P., & Jangir, N. (2017). Non-dominated sorting whale optimization algorithm (NSWOA): A multi-objective optimization algorithm for solving engineering design problems. *Global Journal of Research In Engineering*.
26. Xu, Z., Yu, Y., Yachi, H., Ji, J., Todo, Y., & Gao, S. (2018). A novel memetic whale optimization algorithm for optimization. In *International Conference on Swarm Intelligence* (pp. 384–396). Cham: Springer.
27. Trivedi, I. N., Jangir, P., Kumar, A., Jangir, N., & Totlani, R. (2018). A novel hybrid PSOWOA algorithm for global numerical functions optimization. In *Advances in Computer and Computational Sciences* (pp. 53–60). Singapore: Springer.
28. Kumar, N., Hussain, I., Singh, B., & Panigrahi, B. K. (2017). MPPT in dynamic condition of partially shaded PV system by using WODE technique. *IEEE Transactions on Sustainable Energy*, *8*(3), 1204–1214.
29. Jadhav, A. N., & Gomathi, N. (2017). WGC: Hybridization of exponential grey wolf optimizer with whale optimization for data clustering. *Alexandria Engineering Journal*.

30. Mohamed, F., AbdelNasser, M., Mahmoud, K., & Kamel, S. (2017). Accurate economic dispatch solution using hybrid whale-wolf optimization method. In *2017 Nineteenth International Middle East Power Systems Conference (MEPCON)* (pp. 922–927). IEEE.
31. Singh, N., & Hachimi, H. (2018). A new hybrid whale optimizer algorithm with mean strategy of grey wolf optimizer for global optimization. *Mathematical and Computational Applications, 23*(1), 14.
32. Kaveh, A., & Rastegar Moghaddam, M. (2018). A hybrid WOA-CBO algorithm for construction site layout planning problem. *Scientia Iranica, 25*(3), 1094–1104.
33. Khalilpourazari, S., & Khalilpourazary, S. (2018). SCWOA: An efficient hybrid algorithm for parameter optimization of multi-pass milling process. *Journal of Industrial and Production Engineering, 35*(3), 135–147.
34. Sai, L., & Huajing, F. (2017). A WOA-based algorithm for parameter optimization of support vector regression and its application to condition prognostics. In *2017 36th Chinese Control Conference (CCC)* (pp. 7345–7350). IEEE.
35. Bhesdadiya, R., Jangir, P., Jangir, N., Trivedi, I. N., & Ladumor, D. (2016). Training multilayer perceptron in neural network using whale optimization algorithm. *Indian Journal of Science and Technology, 9*(19), 28–36.
36. Lai, K. H., Zainuddin, Z., & Ong, P. (2017). A study on the performance comparison of metaheuristic algorithms on the learning of neural networks. In *AIP Conference Proceedings* (Vol. 1870, No. 1, p. 040039). AIP Publishing.
37. Yadav, H., Lithore, U., & Agrawal, N. (2017). An enhancement of whale optimization algorithm using ANN for routing optimization in Ad-hoc network. *International Journal of Advanced Technology and Engineering Exploration, 4*(36), 161–167.
38. Chen, Y., Vepa, R., & Shaheed, M. H. (2018). Enhanced and speedy energy extraction from a scaled-up pressure retarded osmosis process with a whale optimization based maximum power point tracking. *Energy, 153*, 618–627.
39. Mehne, H. H., & Mirjalili, S. (2018). A parallel numerical method for solving optimal control problems based on whale optimization algorithm. *Knowledge-Based Systems, 151*, 114–123.
40. Saha, A., & Saikia, L. C. (2018). Performance analysis of combination of ultra-capacitor and superconducting magnetic energy storage in a thermal-gas AGC system with utilization of whale optimization algorithm optimized cascade controller. *Journal of Renewable and Sustainable Energy, 10*(1), 014103.
41. Algabalawy, M. A., Abdelaziz, A. Y., Mekhamer, S. F., & Aleem, S. H. A. (2018). Considerations on optimal design of hybrid power generation systems using whale and sine cosine optimization algorithms. *Journal of Electrical Systems and Information Technology.*
42. Hasanien, H. M. (2018). Performance improvement of photovoltaic power systems using an optimal control strategy based on whale optimization algorithm. *Electric Power Systems Research, 157*, 168–176.
43. Sarkar, P., Laskar, N. M., Nath, S., Baishnab, K. L., & Chanda, S. (2018). Offset voltage minimization based circuit sizing of CMOS operational amplifier using whale optimization algorithm. *Journal of Information and Optimization Sciences, 39*(1), 83–98.
44. Parambanchary, D., & Rao, V. M. WOA-NN: a decision algorithm for vertical handover in heterogeneous networks. *Wireless Networks 1–16.*
45. Jadhav, A. R., & Shankar, T. (2017). Whale optimization based energy-efficient cluster head selection algorithm for wireless sensor networks. arXiv:1711.09389.
46. Kumawat, I. R., Nanda, S. J., & Maddila, R. K. (2018). Positioning LED panel for uniform illuminance in indoor VLC system using whale optimization. In *Optical and Wireless Technologies* (pp. 131–139). Singapore: Springer.
47. Alomari, A., Phillips, W., Aslam, N., & Comeau, F. (2018). Swarm intelligence optimization techniques for obstacle-avoidance mobility-assisted localization in wireless sensor networks. *IEEE Access, 6*, 22368–22385.
48. Rewadkar, D., & Doye, D. (2018). Multiobjective autoregressive whale optimization for traffic-aware routing in urban VANET. *IET Information Security.*

49. Yadav, H., Lilhore, U., & Agrawal, N. (2017). A survey: whale optimization algorithm for route optimization problems. *Wireless Communication, 9*(5), 105–108.
50. Rewadkar, D., & Doye, D. (2018). Adaptive-ARW: adaptive autoregressive whale optimization algorithm for traffic-aware routing in urban VANET. *International Journal of Computer Sciences and Engineering, 6*(2), 303–312.
51. Sharma, M., & Garg, R. (2017). Energy-aware whale-optmized task scheduler in cloud computing. In *2017 International Conference on Intelligent Sustainable Systems (ICISS)* (pp. 121–126). IEEE.
52. Reddy, G. N., & Kumar, S. P. (2017). Multi objective task scheduling algorithm for cloud computing using whale optimization technique. In *International Conference on Next Generation Computing Technologies* (pp. 286–297). Singapore: Springer.
53. Sreenu, K., & Sreelatha, M. (2017). W-Scheduler: whale optimization for task scheduling in cloud computing. *Cluster Computing*, 1–12.
54. Mousavirad, S. J., & Ebrahimpour-Komleh, H. (2017). Multilevel image thresholding using entropy of histogram and recently developed population-based metaheuristic algorithms. *Evolutionary Intelligence, 10*(1–2), 45–75.
55. Sahlol, A. T., & Hassanien, A. E. (2017). Bio-inspired optimization algorithms for Arabic handwritten characters. In *Handbook of research on machine learning innovations and trends* (pp. 897–914). IGI Global.
56. Hassan, G., & Hassanien, A. E. (2018). Retinal fundus vasculature multilevel segmentation using whale optimization algorithm. *Signal, Image and Video Processing, 12*(2), 263–270.
57. Hassanien, A. E., Elfattah, M. A., Aboulenin, S., Schaefer, G., Zhu, S. Y., & Korovin, I. (2016). Historic handwritten manuscript binarisation using whale optimisation. In *2016 IEEE International Conference on Systems, Man, and Cybernetics (SMC)* (pp. 003842–003846). IEEE.
58. Zhang, C., Fu, X., Peng, S., & Wang, Y. (2018). Linear unequally spaced array synthesis for sidelobe suppression with different aperture constraints using whale optimization algorithm. In *2018 13th IEEE Conference on Industrial Electronics and Applications (ICIEA)*. IEEE.
59. Yuan, P., Guo, C., Zheng, Q., & Ding, J. (2018). Sidelobe suppression with constraint for MIMO radar via chaotic whale optimisation. *Electronics Letters, 54*(5), 311–313.
60. Pathak, V. K., & Singh, A. K. (2017). Accuracy control of contactless laser sensor system using whale optimization algorithm and moth-flame optimization. *tm-Technisches Messen, 84*(11), 734–746.
61. Reddy, A. S., & Reddy, M. D. (2018). Application of whale optimization algorithm for distribution feeder reconfiguration. *Journal on Electrical Engineering, 11*(3).
62. Zhang, C., Fu, X., Ligthart, L. P., Peng, S., & Xie, M. (2018). Synthesis of broadside linear aperiodic arrays with sidelobe suppression and null steering using whale optimization algorithm. *IEEE Antennas and Wireless Propagation Letters, 17*(2), 347–350.
63. Yuan, P., Guo, C., Ding, J., & Qu, Y. (2017). Synthesis of nonuniform sparse linear array antenna using whale optimization algorithm. In *2017 Sixth Asia-Pacific Conference on Antennas and Propagation (APCAP)* (pp. 1–3). IEEE.
64. Nasiri, J., & Khiyabani, F. M. (2018). A whale optimization algorithm (WOA) approach for clustering. *Cogent Mathematics & Statistics*, 1483565.
65. Emmanuel, W. S., & Minija, S. J. (2018). Fuzzy clustering and Whale-based neural network to food recognition and calorie estimation for daily dietary assessment. *Sādhanā, 43*(5), 78.
66. Al-Janabi, T. A., & Al-Raweshidy, H. S. (2017). Efficient whale optimisation algorithm-based SDN clustering for IoT focused on node density. In *2017 16th Annual Mediterranean Ad Hoc Networking Workshop (Med-Hoc-Net)* (pp. 1–6). IEEE.
67. Osama, S., Darwish, A., Houssein, E. H., Hassanien, A. E., Fahmy, A. A., & Mahrous, A. (2017). Long-term wind speed prediction based on optimized support vector regression. In *2017 Eighth International Conference on Intelligent Computing and Information Systems (ICICIS)* (pp. 191–196). IEEE.
68. Barham, R., & Aljarah, I. (2017). Link Prediction Based on Whale Optimization Algorithm. In *2017 International Conference on New Trends in Computing Sciences (ICTCS)* (pp. 55–60). IEEE.

69. Osama, S., Houssein, E. H., Hassanien, A. E., & Fahmy, A. A. (2017). Forecast of wind speed based on whale optimization algorithm. In *Proceedings of the 1st International Conference on Internet of Things and Machine Learning* (p. 62). ACM.

70. Desuky, A. S. (2017). Two enhancement levels for male fertility rate categorization using whale optimization and pegasos algorithms. *Australian Journal of Basic and Applied Sciences, 11*(7), 78–83.

71. Sherin, B. M., & Supriya, M. H. (2017). WOA based selection and parameter optimization of SVM kernel function for underwater target classification. *International Journal of Advanced Research in Computer Science, 8*(3).

72. Elazab, O. S., Hasanien, H. M., Elgendy, M. A., & Abdeen, A. M. (2017). Whale optimisation algorithm for photovoltaic model identification. *The Journal of Engineering, 2017*(13), 1906–1911.

73. Yan, Z., Sha, J., Liu, B., Tian, W., & Lu, J. (2018). An ameliorative whale optimization algorithm for multi-objective optimal allocation of water resources in Handan, China. *Water, 10*(1), 87.

74. AlaM, A. Z., Faris, H., & Hassonah, M. A. (2018). Evolving support vector machines using whale optimization algorithm for spam profiles detection on online social networks in different lingual contexts. *Knowledge-Based Systems, 153*, 91–104.

75. Hegazy, A. E., Makhlouf, M. A., & El-Tawel, G. S. (2018). Dimensionality reduction using an improved whale optimization algorithm for data classification. *International Journal of Modern Education and Computer Science, 10*(7), 37.

76. Zamani, H., & Nadimi-Shahraki, M. H. (2016). Feature selection based on whale optimization algorithm for diseases diagnosis. *International Journal of Computer Science and Information Security, 14*(9), 1243.

77. Mafarja, M., & Mirjalili, S. (2018). Whale optimization approaches for wrapper feature selection. *Applied Soft Computing, 62*, 441–453.

78. Ruiye, J., Tao, C., Songyan, W., & Ming, Y. (2018, March). Order whale optimization algorithm in rendezvous orbit design. In *2018 Tenth International Conference on Advanced Computational Intelligence (ICACI)* (pp. 97–102). IEEE.

79. Canayaz, M., & Demir, M. (2017). Feature selection with the whale optimization algorithm and artificial neural network. In *2017 International Artificial Intelligence and Data Processing Symposium (IDAP)* (pp. 1—5). IEEE.

80. Sharawi, M., Zawbaa, H. M., & Emary, E. (2017). Feature selection approach based on whale optimization algorithm. In *2017 Ninth International Conference on Advanced Computational Intelligence (ICACI)* (pp. 163–168). IEEE.

81. Yu, Y., Wang, H., Li, N., Su, Z., & Wu, J. (2017). Automatic carrier landing system based on active disturbance rejection control with a novel parameters optimizer. *Aerospace Science and Technology, 69*, 149–160.

82. Saidala, R. K., & Devarakonda, N. (2018). Improved whale optimization algorithm case study: Clinical data of anaemic pregnant woman. In *Data engineering and intelligent computing* (pp. 271–281). Singapore: Springer.

83. Sayed, G. I., Darwish, A., Hassanien, A. E., & Pan, J. S. (2016). Breast cancer diagnosis approach based on meta-heuristic optimization algorithm inspired by the bubble-net hunting strategy of whales. In *International Conference on Genetic and Evolutionary Computing* (pp. 306–313). Cham: Springer.

84. Mostafa, A., Hassanien, A. E., Houseni, M., & Hefny, H. (2017). Liver segmentation in MRI images based on whale optimization algorithm. *Multimedia Tools and Applications, 76*(23), 24931–24954.

85. Tharwat, A., Moemen, Y. S., & Hassanien, A. E. (2017). Classification of toxicity effects of biotransformed hepatic drugs using whale optimized support vector machines. *Journal of biomedical informatics, 68*, 132–149.

86. Wu, J., Wang, H., Li, N., Yao, P., Huang, Y., & Yang, H. (2018). Path planning for solar-powered UAV in urban environment. *Neurocomputing, 275*, 2055–2065.

87. Dao, T. K., Pan, T. S., & Pan, J. S. (2016). A multi-objective optimal mobile robot path planning based on whale optimization algorithm. In *2016 IEEE 13th International Conference on Signal Processing (ICSP)* (pp. 337–342). IEEE.
88. Kaveh, A., & Ghazaan, M. I. (2017). Enhanced whale optimization algorithm for sizing optimization of skeletal structures. *Mechanics Based Design of Structures and Machines, 45*(3), 345–362.
89. Kaveh, A. (2017). Sizing optimization of skeletal structures using the enhanced whale optimization algorithm. In *Applications of metaheuristic optimization algorithms in civil engineering* (pp. 47–69). Cham: Springer.
90. Prakash, D. B., & Lakshminarayana, C. (2017). Optimal siting of capacitors in radial distribution network using whale optimization algorithm. *Alexandria Engineering Journal, 56*(4), 499–509.
91. Kumar, A., Bhalla, V., Kumar, P., Bhardwaj, T., & Jangir, N. (2018). Whale optimization algorithm for constrained economic load dispatch problems-A cost optimization. In *Ambient Communications and Computer Systems* (pp. 353–366). Singapore: Springer.
92. Khalilpourazari, S., Pasandideh, S. H. R., & Ghodratnama, A. (2019). Robust possibilistic programming for multi-item EOQ model with defective supply batches: Whale optimization and water cycle algorithms. *Neural Computing and Applications*, (in-press).
93. Zhang, X., Liu, Z., Miao, Q., & Wang, L. (2018). Bearing fault diagnosis using a whale optimization algorithm-optimized orthogonal matching pursuit with a combined timefrequency atom dictionary. *Mechanical Systems and Signal Processing, 107*, 29–42.
94. Abdel-Basset, M., Manogaran, G., El-Shahat, D., & Mirjalili, S. (2018). A hybrid whale optimization algorithm based on local search strategy for the permutation flow shop scheduling problem. *Future Generation Computer Systems, 85*, 129–145.
95. Abdel-Basset, M., Manogaran, G., Abdel-Fatah, L., & Mirjalili, S. (2018). An improved nature inspired meta-heuristic algorithm for 1-D bin packing problems. *Personal and Ubiquitous Computing*, 1–16.
96. Horng, M. F., Dao, T. K., & Shieh, C. S. (2017). A Multi-objective optimal vehicle fuel consumption based on whale optimization algorithm. In *Advances in Intelligent Information Hiding and Multimedia Signal Processing* (pp. 371–380). Cham: Springer.
97. Masadeh, R., Alzaqebah, A., & Sharieh, A. (2018). Whale optimization algorithm for solving the maximum flow problem. *Journal of Theoretical & Applied Information Technology, 96*(8)
98. Fu, M., Liao, J., Shao, Z., Marko, M., Zhang, Y., Wang, X., et al. (2016). Finely engineered slow light photonic crystal waveguides for efficient wideband wavelength-independent higher-order temporal solitons. *Applied Optics, 55*(14), 3740–3745.
99. Jiang, L., Wu, H., Jia, W., & Li, X. (2013). Optimization of low-loss and wide-band sharp photonic crystal waveguide bends using the genetic algorithm. *Optik-International Journal for Light and Electron Optics, 124*(14), 1721–1725.
100. Mirjalili, S. M., Mirjalili, S., & Mirjalili, S. Z. (2015). How to design photonic crystal LEDs with artificial intelligence techniques. *Electronics Letters, 51*(18), 1437–1439.
101. Safdari, M. J., Mirjalili, S. M., Bianucci, P., & Zhang, X. (2018). Multi-objective optimization framework for designing photonic crystal sensors. *Applied Optics, 57*(8), 1950–1957.
102. Mirjalili, S. M., Merikhi, B., Mirjalili, S. Z., Zoghi, M., & Mirjalili, S. (2017). Multi-objective versus single-objective optimization frameworks for designing photonic crystal filters. *Applied Optics, 56*(34), 9444–9451.
103. Mirjalili, S. M., & Mirjalili, S. Z. (2017). Single-objective optimization framework for designing photonic crystal filters. *Neural Computing Applications, 28*(6), 1463–1469.

Printed in the United States
By Bookmasters